Illustrated Electrical Calculations

Illustrated Electrical Calculations

Melvin K. Sanders   Ray C. Mullin

Delmar Publishers Inc.

NOTICE TO THE READER

Publisher does not warrant or guarantee any of the products described herein or perform any independent analysis in connection with any of the product information contained herein. Publisher does not assume, and expressly disclaims, any obligation to obtain and include information other than that provided to it by the manufacturer.

The reader is expressly warned to consider and adopt all safety precautions that might be indicated by the activities described herein and to avoid all potential hazards. By following the instructions contained herein, the reader willingly assumes all risks in connection with such instructions.

The publisher makes no representations or warranties of any kind, including but not limited to, the warranties of fitness for particular purpose or merchantability, nor are any such representations implied with respect to the material set forth herein, and the publisher takes no responsibility with respect to such material. The publisher shall not be liable for any special, consequential or exemplary damages resulting, in whole or in part, from the readers' use of, or reliance upon, this material.

Cover design: design M design W

Delmar Staff
Electrical Editor: Mark Huth
Assistant Editor: Nancy Belser
Project Editor: Mary P. Robinson
Production Supervisor: Wendy Troeger
Art/Design Supervisor: Susan M. Mathews

For information, address Delmar Publishers Inc.
3 Columbia Circle
Box 15015
Albany, NY 12212-5015

Printed in the United States of America
Published simultaneously in Canada
by Nelson Canada,
a division of the Thomson Corporation

10 9 8 7 6 5 4 3 2 1 xxx 99 98 97 96 95 94 93

ISBN 0-8273-5462-2

Library of Congress Cataloging-in-Publication Data

Sanders, Melvin K.
Illustrated electrical calculations / Melvin K. Sanders and Ray C. Mullin.
p. cm.
Includes index.
ISBN 0-8273-5462-2
1. Electric engineering—Mathematics—Problems, exercises, etc.
I. Mullin, Ray C. II. Title.
TK153.S323 1993
621′.076—dc20 92-36003
CIP

CONTENTS

PREFACE

The purpose of this book is to provide electricians and those in the electrical industry a guide to the application of recommended concepts when determining circuit requirements. We assume the reader has a working knowledge of mathematical concepts, including basic trigonometry, basic DC and AC principles, and the use and application of electrical standards and codes such as those published, promoted, or used by various association, trade, and engineering organizations.

This book is not intended to instruct students in the theory of the examples and problems contained herein. Explanatory material and definitions are brief, and references to the *National Electrical Code*® are limited to specific sections and tables when it is desired to illustrate a design step. Because it is intended that good design practices are illustrated, no attempt has been made to adhere to code minimums or maximums, but rather to discuss circuit arrangements in terms of "real life" and allow for extra capacity.

Some things to be aware of as you use this book: all sections and tables are based on the 1993 edition of the *National Electrical Code*®; decimals are used to separate numbers equal to or larger than zero from those less than zero, and the practice of using commas to divide numbers into groups of three is continued (although the engineering community is eliminating them because some computer programs read commas as decimals). All math values are rounded up to the next whole number from 0.5 where major fractions are encountered (except for raceway sizing, ohmic values, or other instances in which more significant places must be used). All phase (ungrounded) conductors subject to harmonic currents are increased by 1.25; neutral conductors of linear, single-phase sources are increased by 1.25, and neutral conductors of wye sources are increased by 1.73 for the same reason. Affected equipment, such as panel board, switchboard assemblies, and transformers, are handled the same way. The authors understand the phase conductor and linear neutral source multiplier could easily be 1.0 to 2.0 (or anything in between), and 1.0 to 3.0 for wye source neutral conductors (or anything in between), with transformers sized around various k-factors. The method used here should adequately address these situations until more exact information is available.

It is understood that many approaches and therefore answers could be obtained for the problems and examples given, and if your answer differs from ours, just review the procedures so you are satisfied. After all, that is the purpose behind this book.

The authors and the publisher wish to thank the following experts for reviewing the manuscript and providing constructive suggestions to ensure that this book is as good as it can be:

James Guthrie
John Cox
Dean Blevins
Thomas J. Ritchie, Jr.

The authors wish to acknowledge the assistance of Mark Huth and others at Delmar through the trying times of turning a "wanna-be" into a "can-be."

CHAPTER 1

Direct Current Formulas

1.1 SERIES CIRCUITS

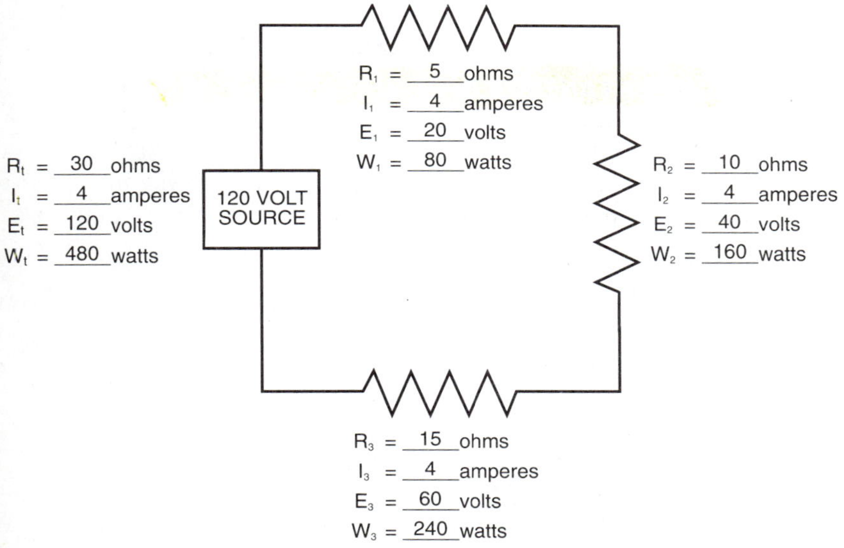

1.1.1 Calculating Watts

All calculations below refer to the figure above.

WATTS = AMPERES × VOLTS **($W = I \times E$)**

EXAMPLE 1:

$W_1 = I_1 \times E_1 = 4 \times 20 = 80$ WATTS

$W_2 = I_2 \times E_2 = 4 \times 40 = 160$ WATTS

$W_3 = I_3 \times E_3 = 4 \times 60 = 240$ WATTS

$W_t = I_t \times E_t = 4 \times 120 = 480$ WATTS

IMPORTANT RULE: $W_t = W_1 + W_2 + W_3 + \ldots$

PRACTICE EXERCISE 1: Determine the wattage in a 120-volt circuit when 10 amperes are flowing.

WATTS = AMPERES² × OHMS $(W = I^2 \times R)$

EXAMPLE 2:

$$W_1 = I_1^2 \times R_1 = 4 \times 4 \times 5 = 80 \text{ WATTS}$$
$$W_2 = I_2^2 \times R_2 = 4 \times 4 \times 10 = 160 \text{ WATTS}$$
$$W_3 = I_3^2 \times R_3 = 4 \times 4 \times 15 = 240 \text{ WATTS}$$
$$W_t = I_t^2 \times R_t = 4 \times 4 \times 30 = 480 \text{ WATTS}$$

IMPORTANT RULE: $W_t = W_1 + W_2 + W_3 + \ldots$

PRACTICE EXERCISE 2: Find the watts loss in the conductors having a resistance of 0.5 ohms. The current draw is 25 amperes.

$$\textbf{WATTS} = \frac{\textbf{VOLTS}^2}{\textbf{OHMS}} \qquad \left(W = \frac{E^2}{R}\right)$$

EXAMPLE 3:

$$W_1 = \frac{E_1^2}{R_1} = \frac{20 \times 20}{5} = 80 \text{ WATTS}$$
$$W_2 = \frac{E_2^2}{R_2} = \frac{40 \times 40}{10} = 160 \text{ WATTS}$$
$$W_3 = \frac{E_3^2}{R_3} = \frac{60 \times 60}{15} = 240 \text{ WATTS}$$
$$W_t = \frac{E_t^2}{R_t} = \frac{120 \times 120}{30} = 480 \text{ WATTS}$$

IMPORTANT RULE: $W_t = W_1 + W_2 + W_3 + \ldots$

PRACTICE EXERCISE 3: The voltage across the line terminals of a baseboard resistance electric heater is 240 volts. The heating element is found to have a hot resistance of 23 ohms. Calculate the wattage.

1.1.2 Calculating Volts

All calculations below refer to the figure on page 1.

$$\textbf{VOLTS = AMPERES} \times \textbf{OHMS} \qquad (\mathbf{E = I \times R})$$

EXAMPLE 1:
$$E_1 = I_1 \times R_1 = 4 \times 5 = 20\text{ VOLTS}$$
$$E_2 = I_2 \times R_2 = 4 \times 10 = 40\text{ VOLTS}$$
$$E_3 = I_3 \times R_3 = 4 \times 15 = 60\text{ VOLTS}$$
$$E_t = I_t \times R_t = 4 \times 30 = 120\text{ VOLTS}$$

> **IMPORTANT RULE for series circuits:** $E_t = E_1 + E_2 + E_3 + \ldots$

PRACTICE EXERCISE 1: The current flowing through a 7.5 ohm resistor measures 16 amperes. What is the circuit voltage?

___

$$\textbf{VOLTS} = \sqrt{\textbf{WATTS} \times \textbf{OHMS}} \qquad (\mathbf{E} = \sqrt{\mathbf{W \times R}})$$

EXAMPLE 2:
$$E_1 = \sqrt{W_1 \times R_1} = \sqrt{80 \times 5} = \sqrt{400} = 20\text{ VOLTS}$$
$$E_2 = \sqrt{W_2 \times R_2} = \sqrt{160 \times 10} = \sqrt{1{,}600} = 40\text{ VOLTS}$$
$$E_3 = \sqrt{W_3 \times R_3} = \sqrt{240 \times 15} = \sqrt{3{,}600} = 60\text{ VOLTS}$$
$$E_t = \sqrt{W_t \times R_t} = \sqrt{480 \times 30} = \sqrt{14{,}400} = 120\text{ VOLTS}$$

> **IMPORTANT RULE:** $E_t = E_1 + E_2 + E_3 + \ldots$

PRACTICE EXERCISE 2: The nameplate on a heater indicates 5,000 watts. The resistance of the heating element is found to be 8.653 ohms. What voltage must be applied to the heating element to produce 5,000 watts?

___

$$\textbf{VOLTS} = \frac{\textbf{WATTS}}{\textbf{AMPERES}} \qquad \left(\mathbf{E} = \frac{\mathbf{W}}{\mathbf{I}}\right)$$

EXAMPLE 3:

$$E_1 = \frac{W_1}{I_1} = \frac{80}{4} = 20 \text{ VOLTS}$$

$$E_2 = \frac{W_2}{I_2} = \frac{160}{4} = 40 \text{ VOLTS}$$

$$E_3 = \frac{W_3}{I_3} = \frac{240}{4} = 60 \text{ VOLTS}$$

$$E_t = \frac{W_t}{I_t} = \frac{480}{4} = 120 \text{ VOLTS}$$

IMPORTANT RULE: $E_t = E_1 + E_2 + E_3 + \ldots$

PRACTICE EXERCISE 3: The nameplate of an appliance shows 3,000 watts, 25 amperes. The voltage marking cannot be read. Is this appliance intended to be connected to 120 volts or 240 volts?

1.1.3 Calculating Ohms

All calculations below refer to the figure on page 1.

$$\text{OHMS} = \frac{\text{VOLTS}}{\text{AMPERES}} \qquad \left(R = \frac{E}{I}\right)$$

EXAMPLE 1:

$$R_1 = \frac{E_1}{I_1} = \frac{20}{4} = 5 \text{ OHMS}$$

$$R_2 = \frac{E_2}{I_2} = \frac{40}{4} = 10 \text{ OHMS}$$

$$R_3 = \frac{E_3}{I_3} = \frac{60}{4} = 15 \text{ OHMS}$$

$$R_t = \frac{E_t}{I_t} = \frac{120}{4} = 30 \text{ OHMS}$$

IMPORTANT RULE: $R_t = R_1 + R_2 + R_3 + \ldots$

PRACTICE EXERCISE 1: Find the resistance of a circuit rated 277 volts when the current draw is 12 amperes.

$$\textbf{OHMS} = \frac{\textbf{VOLTS}^2}{\textbf{WATTS}} \qquad \left(\mathbf{R} = \frac{\mathbf{E}^2}{\mathbf{W}}\right)$$

EXAMPLE 2:

$$R_1 = \frac{E_1^2}{W_1} = \frac{20 \times 20}{80} = 5 \text{ OHMS}$$

$$R_2 = \frac{E_2^2}{W_2} = \frac{40 \times 40}{160} = 10 \text{ OHMS}$$

$$R_3 = \frac{E_3^2}{W_3} = \frac{60 \times 60}{240} = 15 \text{ OHMS}$$

$$R_t = \frac{E_t^2}{W_t} = \frac{120 \times 120}{480} = 30 \text{ OHMS}$$

IMPORTANT RULE: $R_t = R_1 + R_2 + R_3 + \ldots$

PRACTICE EXERCISE 2: Calculate the resistance of a circuit rated 277 volts where the wattage is 3,324 watts.

$$\textbf{OHMS} = \frac{\textbf{WATTS}}{\textbf{AMPERES}^2} \qquad \left(\mathbf{R} = \frac{\mathbf{W}}{\mathbf{I}^2}\right)$$

EXAMPLE 3:

$$R_1 = \frac{W_1}{I_1^2} = \frac{80}{4 \times 4} = 5 \text{ OHMS}$$

$$R_2 = \frac{W_2}{I_2^2} = \frac{160}{4 \times 4} = 10 \text{ OHMS}$$

$$R_3 = \frac{W_3}{I_3^2} = \frac{240}{4 \times 4} = 15 \text{ OHMS}$$

$$R_t = \frac{W_t}{I_t^2} = \frac{480}{4 \times 4} = 30 \text{ OHMS}$$

IMPORTANT RULE: $R_t = R_1 + R_2 + R_3 + \ldots$

PRACTICE EXERCISE 3: A 1,500-watt load, when properly connected, draws 10 amperes. Find the resistance.

1.1.4 Calculating Amperes

All calculations below refer to the figure on page 1.

$$\textbf{AMPERES} = \frac{\textbf{VOLTS}}{\textbf{OHMS}} \qquad \left(\mathbf{I} = \frac{\mathbf{E}}{\mathbf{R}}\right)$$

EXAMPLE 1:

$$I_1 = \frac{E_1}{R_1} = \frac{20}{5} = 4 \text{ AMPERES}$$

$$I_2 = \frac{E_2}{R_2} = \frac{40}{10} = 4 \text{ AMPERES}$$

$$I_3 = \frac{E_3}{R_3} = \frac{60}{15} = 4 \text{ AMPERES}$$

$$I_t = \frac{E_t}{R_t} = \frac{120}{30} = 4 \text{ AMPERES}$$

> **IMPORTANT RULE:** $I_t = I_1 = I_2 = I_3 \ldots$

PRACTICE EXERCISE 1: A conductor that has a resistance of 0.1 ohms is shorted across a 240-volt source. How much current will flow in this short circuit?

$$\textbf{AMPERES} = \frac{\textbf{WATTS}}{\textbf{VOLTS}} \qquad \left(\mathbf{I} = \frac{\mathbf{W}}{\mathbf{E}}\right)$$

EXAMPLE 2:

$$I_1 = \frac{W_1}{E_1} = \frac{80}{20} = 4 \text{ AMPERES}$$

$$I_2 = \frac{W_2}{E_2} = \frac{160}{40} = 4 \text{ AMPERES}$$

$$I_3 = \frac{W_3}{E_3} = \frac{240}{60} = 4 \text{ AMPERES}$$

$$I_t = \frac{W_t}{E_t} = \frac{480}{120} = 4 \text{ AMPERES}$$

> **IMPORTANT RULE:** $I_t = I_1 = I_2 = I_3 \ldots$

PRACTICE EXERCISE 2: A toaster is marked 1,000 watts, 120 volts. Calculate the current flow.

$$\text{AMPERES} = \sqrt{\frac{\text{WATTS}}{\text{OHMS}}} \qquad \left(I = \sqrt{\frac{W}{R}}\right)$$

EXAMPLE 3:

$$I_1 = \sqrt{\frac{W_1}{R_1}} = \sqrt{\frac{80}{5}} = \sqrt{16} = 4 \text{ AMPERES}$$

$$I_2 = \sqrt{\frac{W_2}{R_2}} = \sqrt{\frac{160}{10}} = \sqrt{16} = 4 \text{ AMPERES}$$

$$I_3 = \sqrt{\frac{W_3}{R_3}} = \sqrt{\frac{240}{15}} = \sqrt{16} = 4 \text{ AMPERES}$$

$$I_t = \sqrt{\frac{W_t}{R_t}} = \sqrt{\frac{480}{30}} = \sqrt{16} = 4 \text{ AMPERES}$$

IMPORTANT RULE: $I_t = I_1 = I_2 = I_3 \ldots$

PRACTICE EXERCISE 3: Find the current draw of a 1,000-watt lamp having a resistance of 10 ohms.

PRACTICE EXERCISE 4: Solve for all of the unknown values in the diagram.

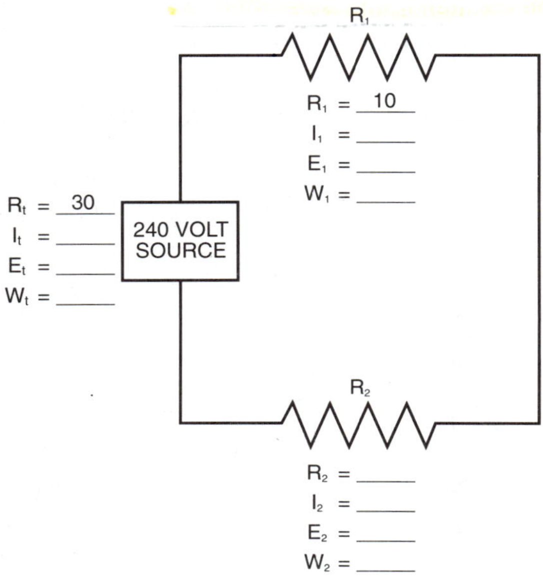

1.2 PARALLEL CIRCUITS

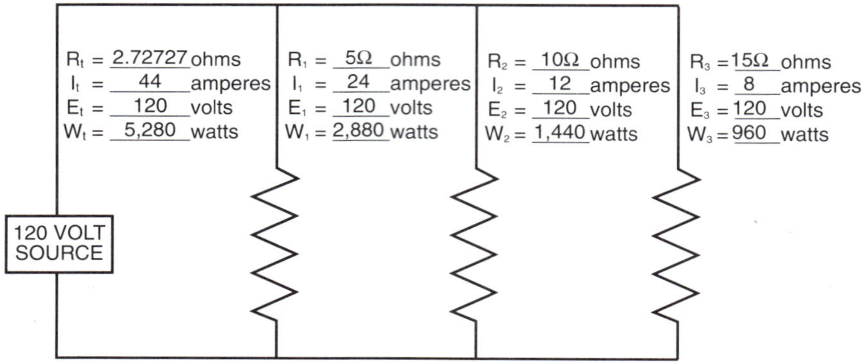

1.2.1 Calculating Ohms

All calculations below refer to the figure above.

> **IMPORTANT RULE:** The total resistance, in ohms, of a parallel circuit is always less than the resistance value of the smallest (lowest resistance) value of resistance of any of the resistors connected in parallel. Sometimes this is termed "smaller than the smallest."

$$\textbf{OHMS} = \frac{\textbf{VOLTS}}{\textbf{AMPERES}} \qquad \left(\mathbf{R} = \frac{\mathbf{E}}{\mathbf{I}}\right)$$

EXAMPLE 1:

$$R_1 = \frac{E_1}{I_1} = \frac{120}{24} = 5 \text{ OHMS}$$

$$R_2 = \frac{E_2}{I_2} = \frac{120}{12} = 10 \text{ OHMS}$$

$$R_3 = \frac{E_3}{I_3} = \frac{120}{8} = 15 \text{ OHMS}$$

$$R_t = \frac{E_t}{I_t} = \frac{120}{44} = 2.7272727 \text{ OHMS}$$

PRACTICE EXERCISE 1: Find the resistance value of a 480-volt, 24-ampere circuit.

> **IMPORTANT RULE:** The voltage across all parts of a parallel circuit is the same.
> $E_t = E_1 = E_2 = E_3 = E_n \ldots$

$$\textbf{OHMS} = \frac{\textbf{VOLTS}^2}{\textbf{WATTS}} \qquad \left(\mathbf{R} = \frac{\mathbf{E}^2}{\mathbf{W}}\right)$$

EXAMPLE 2:

$$R_1 = \frac{E_1^2}{W_1} = \frac{120 \times 120}{2,880} = 5 \text{ OHMS}$$

$$R_2 = \frac{E_2^2}{W_2} = \frac{120 \times 120}{1,440} = 10 \text{ OHMS}$$

$$R_3 = \frac{E_3^2}{W_3} = \frac{120 \times 120}{960} = 15 \text{ OHMS}$$

$$R_t = \frac{E_t^2}{W_t} = \frac{120 \times 120}{5,280} = 2.72727272 \text{ OHMS}$$

PRACTICE EXERCISE 2: Find the resistance value of a 120-volt, 300-watt bulb.

$$\textbf{OHMS} = \frac{\textbf{WATTS}}{\textbf{AMPERES}^2} \qquad \left(\mathbf{R} = \frac{\mathbf{W}}{\mathbf{I}^2}\right)$$

EXAMPLE 3:

$$R_1 = \frac{W_1}{I_1^2} = \frac{2,880}{24 \times 24} = 5 \text{ OHMS}$$

$$R_2 = \frac{W_2}{I_2^2} = \frac{1,440}{12 \times 12} = 10 \text{ OHMS}$$

$$R_3 = \frac{W_3}{I_3^2} = \frac{960}{8 \times 8} = 15 \text{ OHMS}$$

$$R_t = \frac{W_t}{I_t^2} = \frac{5,280}{44 \times 44} = 2.72727272 \text{ OHMS}$$

PRACTICE EXERCISE 3: Find the resistance value of a 1,000-watt flat iron rated 120 volts. Prove this resistance result by using $R = \frac{W}{I^2}$.

Hint: You will need to find amperes first, using watts divided by volts.

1.2.2 Calculating Amperes

All calculations below refer to the figure on page 8.

> **IMPORTANT RULE:** The total current through a parallel circuit is the sum of the currents through each part of the circuit.
>
> $$I_t = I_1 + I_2 + I_3 + I_x \ldots$$

$$\textbf{AMPERES} = \frac{\textbf{VOLTS}}{\textbf{OHMS}} \qquad \left(\mathbf{I} = \frac{\mathbf{E}}{\mathbf{R}}\right)$$

EXAMPLE 1:

$$I_1 = \frac{E_1}{R_1} = \frac{120}{5} = 24 \text{ AMPERES}$$

$$I_2 = \frac{E_2}{R_2} = \frac{120}{10} = 12 \text{ AMPERES}$$

$$I_3 = \frac{E_3}{R_3} = \frac{120}{15} = 8 \text{ AMPERES}$$

$$I_t = \frac{E_t}{R_t} = \frac{120}{2.72727272} = 44 \text{ AMPERES}$$

PRACTICE EXERCISE 1: Given a resistance value of 25 ohms and a voltage of 120 volts, find the current flow.

$$\textbf{AMPERES} = \frac{\textbf{WATTS}}{\textbf{VOLTS}} \qquad \left(\mathbf{I} = \frac{\mathbf{W}}{\mathbf{E}}\right)$$

EXAMPLE 2:

$$I_1 = \frac{W_1}{E_1} = \frac{2{,}880}{120} = 24 \text{ AMPERES}$$

$$I_2 = \frac{W_2}{E_2} = \frac{1{,}440}{120} = 12 \text{ AMPERES}$$

$$I_3 = \frac{W_3}{E_3} = \frac{960}{120} = 8 \text{ AMPERES}$$

$$I_t = \frac{W_t}{E_t} = \frac{5{,}280}{120} = 44 \text{ AMPERES}$$

PRACTICE EXERCISE 2: A heating cable is marked 120 volts, 2,500 watts. Find the cable current draw.

$$\textbf{AMPERES} = \sqrt{\frac{\textbf{WATTS}}{\textbf{OHMS}}} \qquad \left(\mathbf{I} = \sqrt{\frac{\mathbf{W}}{\mathbf{R}}}\right)$$

EXAMPLE 3:

$$I_1 = \sqrt{\frac{W_1}{R_1}} = \sqrt{\frac{2,880}{5}} = \sqrt{576} = 24 \text{ AMPERES}$$

$$I_2 = \sqrt{\frac{W_2}{R_2}} = \sqrt{\frac{1,440}{10}} = \sqrt{144} = 12 \text{ AMPERES}$$

$$I_3 = \sqrt{\frac{W_3}{R_3}} = \sqrt{\frac{960}{15}} = \sqrt{64} = 8 \text{ AMPERES}$$

$$I_t = \sqrt{\frac{W_t}{R_t}} = \sqrt{\frac{5,280}{2.72727272}} = \sqrt{1936} = 44 \text{ AMPERES}$$

PRACTICE EXERCISE 3: Calculate the current draw of a 5,000-watt, 277-volt heating element. Prove that $I = \sqrt{\frac{W}{R}}$.

1.2.3 Calculating Volts

All calculations below refer to the figure on page 8.

$$\textbf{VOLTS} = \textbf{AMPERES} \times \textbf{OHMS} \qquad (\mathbf{E} = \mathbf{I} \times \mathbf{R})$$

EXAMPLE 1:

$$E_1 = I_1 \times R_1 = 24 \times 5 = 120 \text{ VOLTS}$$

$$E_2 = I_2 \times R_2 = 12 \times 10 = 120 \text{ VOLTS}$$

$$E_3 = I_3 \times R_3 = 8 \times 15 = 120 \text{ VOLTS}$$

$$E_t = I_t \times R_t = 44 \times 2.727272 = 120 \text{ VOLTS}$$

> **IMPORTANT RULE:** The voltage across all parts of a parallel circuit is the same.
>
> $E_t = E_1 = E_2 = E_3 = E_x \ldots$

PRACTICE EXERCISE 1: A snow-melting heating cable has a resistance of 15 ohms. It has a 16-ampere rating. What is the correct supply voltage?

$$\textbf{VOLTS} = \frac{\textbf{WATTS}}{\textbf{AMPERES}} \qquad \left(\mathbf{E} = \frac{\mathbf{W}}{\mathbf{I}}\right)$$

EXAMPLE 2:

$$E_1 = \frac{W_1}{I_1} = \frac{2,880}{24} = 120\text{ VOLTS}$$

$$E_2 = \frac{W_2}{I_2} = \frac{1,440}{12} = 120\text{ VOLTS}$$

$$E_3 = \frac{W_3}{I_3} = \frac{960}{8} = 120\text{ VOLTS}$$

$$E_t = \frac{W_t}{I_t} = \frac{5,280}{44} = 120\text{ VOLTS}$$

PRACTICE EXERCISE 2: An appliance is marked 12.5 amperes, 1,500 watts. The voltage rating is illegible. What is the intended voltage supply?

$$\textbf{VOLTS} = \sqrt{\textbf{WATTS} \times \textbf{OHMS}} \quad (\mathbf{E} = \sqrt{\mathbf{W} \times \mathbf{R}})$$

EXAMPLE 3:

$$E_1 = \sqrt{W_1 \times R_1} = \sqrt{2,880 \times 5} = 120\text{ VOLTS}$$

$$E_2 = \sqrt{W_2 \times R_2} = \sqrt{1,440 \times 10} = 120\text{ VOLTS}$$

$$E_3 = \sqrt{W_3 \times R_3} = \sqrt{960 \times 15} = 120\text{ VOLTS}$$

$$E_t = \sqrt{W_t \times R_t} = \sqrt{5,280 \times 2.727272} = 120\text{ VOLTS}$$

PRACTICE EXERCISE 3: An electric furnace is marked 15 kW (kilowatts). An ohmmeter reading shows the resistance to be approximately 2.88 ohms. To produce 15 kW of heat, what voltage must the source voltage be?

1.2. 4 Calculating Watts

All calculations below refer to the figure on page 8.

$$\textbf{WATTS} = \textbf{VOLTS} \times \textbf{AMPERES} \qquad (\mathbf{W} = \mathbf{E} \times \mathbf{I})$$

EXAMPLE 1:
$$W_1 = E_1 \times I_1 = 120 \times 24 = 2{,}880 \text{ WATTS}$$
$$W_2 = E_2 \times I_2 = 120 \times 12 = 1{,}440 \text{ WATTS}$$
$$W_3 = E_3 \times I_3 = 120 \times 8 = 960 \text{ WATTS}$$
$$W_t = E_t \times I_t = 120 \times 44 = 5{,}280 \text{ WATTS}$$

IMPORTANT RULE: $W_t = W_1 + W_2 + W_3 + \ldots$

PRACTICE EXERCISE 1: A 277-volt resistance type duct electric heater is marked 10.83 amperes. Calculate the wattage.

WATTS = AMPERES² × OHMS **($W = I^2 \times R$)**

EXAMPLE 2:
$$W_1 = I_1^2 R_1 = 24 \times 24 \times 5 = 2{,}880 \text{ WATTS}$$
$$W_2 = I_2^2 R_2 = 12 \times 12 \times 10 = 1{,}440 \text{ WATTS}$$
$$W_3 = I_3^2 R_3 = 8 \times 8 \times 15 = 960 \text{ WATTS}$$
$$W_t = I_t^2 R_t = 44 \times 44 \times 2.727272 = 5{,}280 \text{ WATTS}$$

PRACTICE EXERCISE 2: A resistive element measures 25.577 ohms. The nameplate indicates 10.83 amperes, 277 volts. Calculate the element wattage.

$$\textbf{WATTS} = \frac{\textbf{VOLTAGE}^2}{\textbf{OHMS}} \qquad \left(W = \frac{E^2}{R}\right)$$

EXAMPLE 3:
$$W_1 = \frac{E_1^2}{R_1} = \frac{120 \times 120}{5} = 2{,}880 \text{ WATTS}$$
$$W_2 = \frac{E_2^2}{R_2} = \frac{120 \times 120}{10} = 1{,}440 \text{ WATTS}$$
$$W_3 = \frac{E_3^2}{R_3} = \frac{120 \times 120}{15} = 960 \text{ WATTS}$$
$$W_t = \frac{E_t^2}{R_t} = \frac{120 \times 120}{2.727272} = 5{,}280 \text{ WATTS}$$

PRACTICE EXERCISE 3: A resistive element measures 25.577 ohms. Find the element's wattage output when connected to a 277-volt circuit.

PRACTICE EXERCISE 4: Solve for all of the unknown values in the diagram.

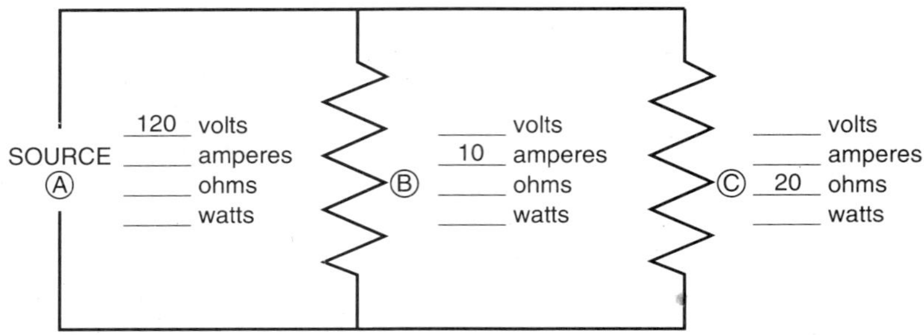

1.2.5 Calculating Ohms Using "Product over the Sum" Method

The following parallel circuit has three resistors in parallel. For such an arrangement, the "PRODUCT OVER THE SUM" method is an easy way to solve the problem. When the circuit has more than two resistors in parallel, take two resistors at a time. Solve for the equivalent resistance of these two resistors. Then use this equivalent resistance value with another resistor and solve again. Keep repeating until the total parallel circuit has been solved.

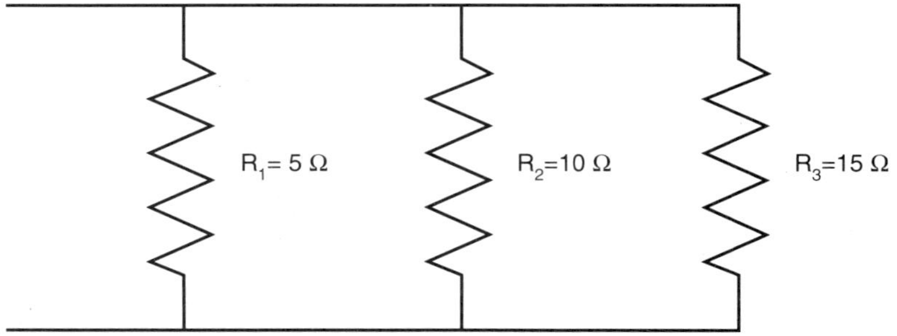

EXAMPLE

Use the R values from the figure above, taking two resistors at a time.

R_t for the 5 and 10.

$$R_t = \frac{R_1 \times R_2}{R_1 + R_2} = \frac{5 \times 10}{5 + 10} = \frac{50}{15} = 3.333333 \text{ OHMS}$$

Now use this result with the 15-ohm resistor.

$$R_t = \frac{3.333333 \times 15}{3.33333 + 15} = \frac{50}{18.333333} = 2.72727 \text{ OHMS}$$

Note that the total resistance of the circuit is "smaller than the smallest."

PRACTICE EXERCISE: Three resistors are connected in parallel. Calculate the circuit's combined resistance using the "PRODUCT OVER THE SUM" method.

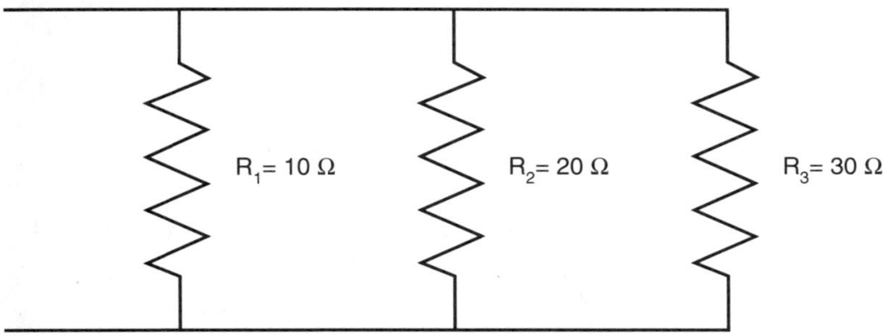

1.2.6 Calculating Total Resistance When All Resistors Are the Same Value

$$R_t = \frac{\text{Resistance in ohms of one resistor}}{\text{Number of resistors connected in parallel}} \qquad \left(R_t = \frac{R_1}{N}\right)$$

EXAMPLE:

Calculate the total circuit resistance when three 25-ohm resistors are connected in parallel.

$$R_t = \frac{R_1}{N} = \frac{25}{3} = 8.33333 \text{ OHMS}$$

PRACTICE EXERCISE: Calculate the total circuit resistance when five 200-ohm resistors are connected in parallel.

1.2.7 Calculating Total Resistance When a Number of Resistors of Various Values Are Connected in Parallel

The most commonly used method for this calculation is referred to as the "RECIPROCAL OF THE SUM OF THE RECIPROCALS." Use of pocket calculators makes this method very easy.

$$\mathbf{R_t = \frac{1}{\frac{1}{R_1} + \frac{1}{R_2} + \frac{1}{R_3}}}$$

EXAMPLE:

Use the resistance values from the figure on the bottom of page 14.

$$R_t = \frac{1}{\frac{1}{5} + \frac{1}{10} + \frac{1}{15}}$$

$$R_t = \frac{1}{0.2 + 0.1 + 0.066666666}$$

$$R_t = \frac{1}{0.366666666}$$

$$R_t = 2.727272727$$

> **IMPORTANT RULE:** The total resistance of a parallel circuit is always less than the lowest value resistor in the circuit.

PRACTICE EXERCISE: A parallel circuit consists of four resistors. Values are 10 ohms, 20 ohms, 40 ohms, and 50 ohms. Find the circuit's combined resistance.

1.3 SERIES/PARALLEL CIRCUITS

To solve for the total resistance of a circuit that is made up of both series and parallel circuits, examine the circuit to determine what parts can be resolved as in series circuits and what parts can be resolved as in parallel circuits. Then combine the parts, step by step, to arrive at the circuit's total resistance. In some cases, the parallel parts must be resolved first before combining. In other cases, the series parts must be totalled first before combining. It is just a matter of analyzing the circuit, then breaking it down into smaller segments.

EXAMPLE 1: In the figure below we must first resolve the parallel portion, then combine that equivalent with the five-ohm resistor to arrive at the circuit's total resistance.

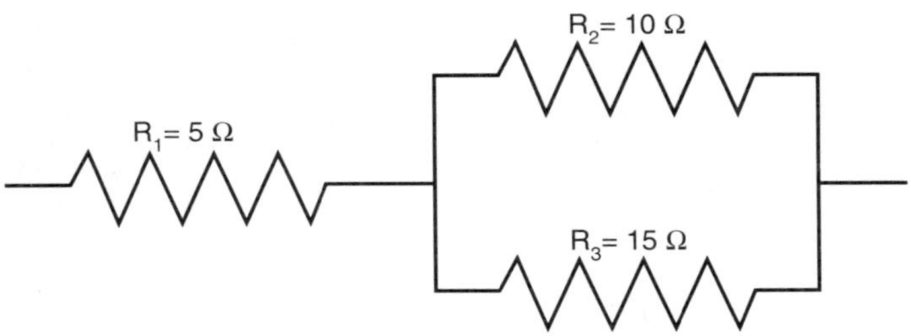

Step 1:
Solve the parallel part of the circuit.

$$\frac{R_2 \times R_3}{R_2 + R_3} = \frac{10 \times 15}{10 + 15} = \frac{150}{25} = 6 \text{ OHMS}$$

Step 2:
The result in Step 1 is in series with R_1. The equivalent circuit is now:

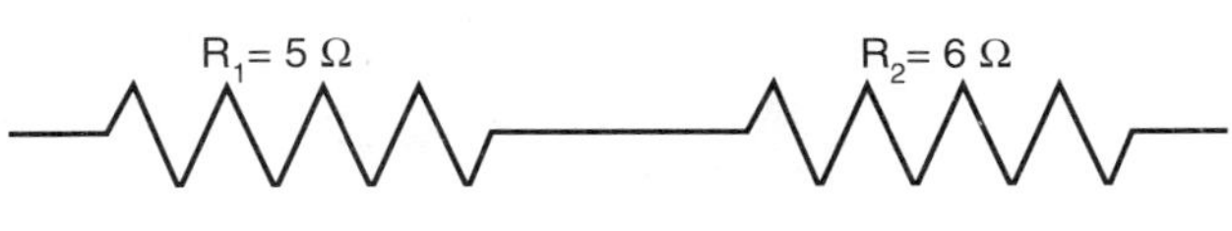

Therefore $R_t = 5 + 6 = 11$ OHMS

PRACTICE EXERCISE 1: Solve for the combined resistance of the following:

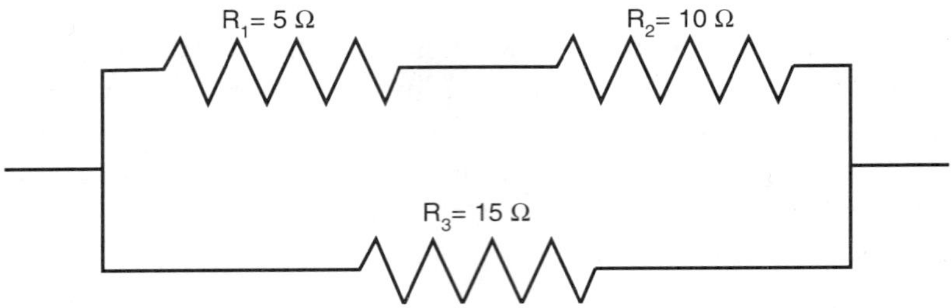

EXAMPLE 2:

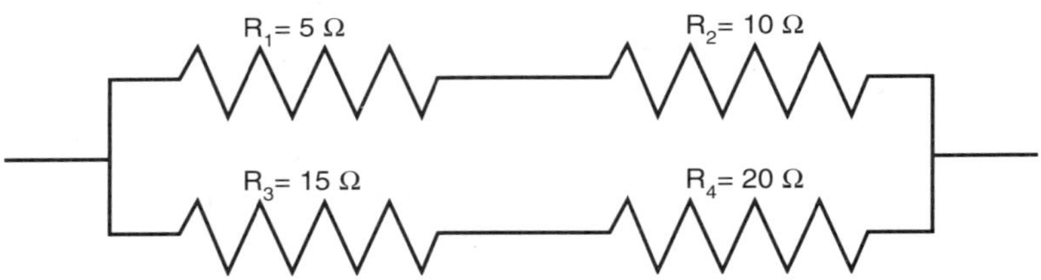

Step 1:

Solve for the series parts of the circuit.

$R_1 + R_2 = 5 + 10 = 15$ OHMS

and $R_3 + R_4 = 15 + 20 = 35$ OHMS

The equivalent circuit is now:

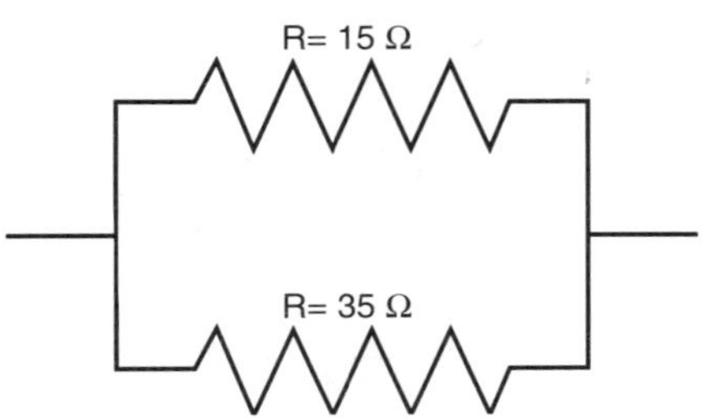

Step 2:

Solve for the equivalent parallel circuit.

$$R_t = \frac{15 \times 35}{15 + 35} = \frac{525}{50} = 10.5 \text{ OHMS}$$

PRACTICE EXERCISE 2: Solve for the combined resistance of the diagram.

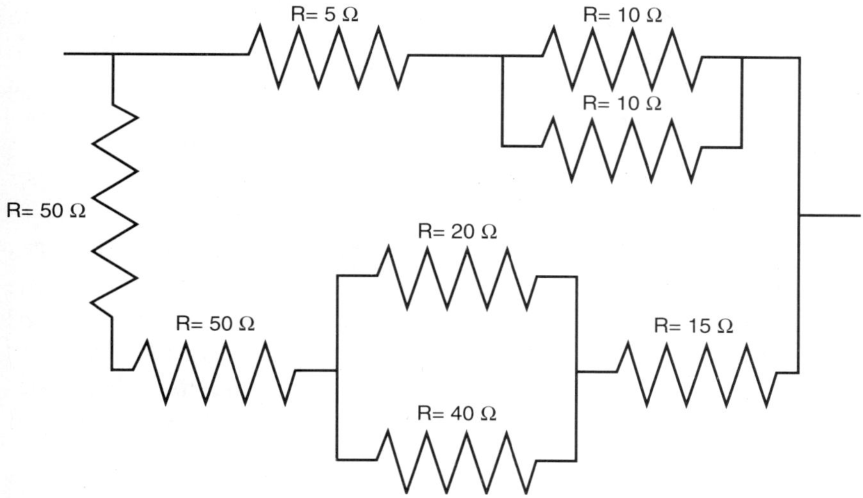

PRACTICE EXERCISE 3: Two 500-kcmil copper conductors are run in parallel. The length of the run is 100 feet. The resistance of each 500-kcmil copper conductor is 0.0258 ohms per 1,000 feet, according to *Table 8* of Chapter 9 in the *National Electrical Code*®. The current flow is 300 amperes. What is the voltage drop across this parallel run?

1.4 CONDUCTORS IN PARALLEL

Section 310-4 of the *National Electrical Code®* requires that when conductors are run in parallel, such as for large services and feeders, the conductors shall 1) be the same length; 2) be the same material; 3) be the same circular mil area; 4) have the same insulation type; and 5) be terminated in the same manner. In some cases, it is virtually impossible to have the same length of conductor because of the manner in which the conductors are routed. There could be a slight difference in length. To make sure that the conductors are not overloaded, the following formula can be used to calculate the division of current through each of the conductors run in parallel per phase.

EXAMPLE:
Three sets of 600-kcmil THW copper conductors are run in parallel. The service is intended to be a 1,200-ampere service. Because of how the conductors are routed through the current transformers and connected to the main bolted pressure switch, the actual lengths turn out to be 20 feet, 22 feet, and 24 feet respectively. The maximum allowable ampacity for the 600-kcmil copper conductors, according to *Table 310-16* of the *National Electrical Code®*, is 420 amperes.

Calculate the current flow in each of the three conductors when the total current per phase is 1,200 amperes.

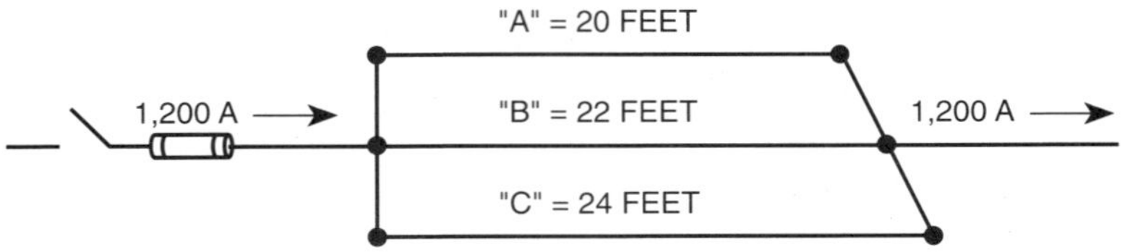

The total length of conductors per phase is:

$$20 + 22 + 24 = 66 \text{ feet}$$

$$\text{Calculation \#1: } \frac{20}{66} \times 1{,}200 = 364 \text{ AMPERES}$$

$$\text{Calculation \#2: } \frac{22}{66} \times 1{,}200 = 400 \text{ AMPERES}$$

$$\text{Calculation \#3: } \frac{24}{66} \times 1{,}200 = 436 \text{ AMPERES}$$

IMPORTANT RULE: Because current flow is inversely proportional to resistance (the higher the resistance, the lower the current), merely list the conductors in reverse order to show that the shortest conductor (the lowest resistance) will share the highest amount of current. The longest conductor (the highest resistance) will share the smallest amount of current. In the above example:

CONDUCTOR	*CURRENT*
A = 20 feet	436 amperes
B = 22 feet	400 amperes
C = 24 feet	364 amperes

Therefore, we can see that Conductor "A" will carry 436 amperes, which exceeds its ampacity of 420 amperes. This does not "meet code," and is a violation of Section 310-4, *National Electrical Code®*.

PRACTICE EXERCISE: Recalculate Example 1 using lengths of 20 feet, 21 feet, and 22 feet for the three 600-kcmil copper conductors run in parallel.

Calculation #1: _____ × 1,200 =_____ AMPERES

Calculation #2: _____ × 1,200 =_____ AMPERES

Calculation #3: _____ × 1,200 =_____ AMPERES

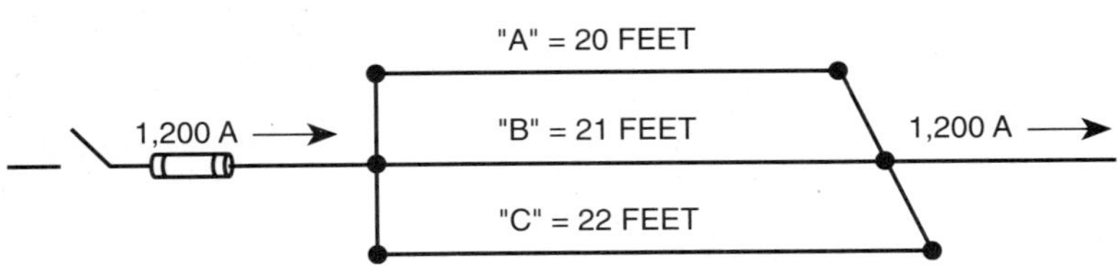

1.5 ENCLOSURES (NEMA)

The *National Electrical Code*® discusses damp locations, dry locations, and wet locations. The National Electrical Manufacturers Association has established a numbering system to identify the suitable use of enclosures.

Enclosure Type Number	Use
1	Indoor use. Provides some protection against limited amount of falling dirt.
2	Indoor use. Provides some protection against limited amount of falling dirt and water.
3	Outdoor use. Provides some protection against rain, sleet, windblown dust, and damage from external ice formation.
3R	Outdoor use. Provides some protection against rain, sleet, and damage from external ice formation.
3S	Outdoor use. Provides some protection against rain, sleet, windblown dust. When laden with ice, the external mechanism remains operable.
4	Indoor or outdoor use. Provides some protection against windblown dust and rain, splashing water, hose-directed water, and damage from external ice formation.
4X	Indoor or outdoor use. Provides some protection against corrosion, windblown dust and rain, splashing water, hose-directed water, and damage from external ice formation.
5	Indoor use. Provides some protection against falling dirt, settling airborne dust, and dripping noncorrosive liquids.
6	Indoor or outdoor use. Provides some protection against hose-directed water, entry of water during occasional temporary submersion at a limited depth, and damage from external ice formation.
6P	Indoor or outdoor use. Provides some protection against hose-directed water, entry of water during prolonged submersion at a limited depth, and damage from external ice formation.
7	Hazardous Class I locations. Indoor use. Groups A, B, C, or D classifications. See *Article 500, National Electrical Code*®.
8	Hazardous Class I locations. Indoor or outdoor use. Groups A, B, C, or D classifications. See *Article 500, National Electrical Code*®.
9	Hazardous Class II locations. Indoor use. Groups E, F, or G classifications. See *Article 500, National Electrical Code*®.

Enclosure Type Number	Use
10	Meets requirements of the Mine Safety and Health Administration (MSHA) regulations.
12	Indoor use. Provides some protection against falling dirt, circulating dust, and dripping noncorrosive liquids.
12K	Indoor use. Has knockouts. Provides some protection against falling dirt, circulating dust, and dripping noncorrosive liquids.
13	Indoor use. Provides some protection against spraying water, dust, oil, and noncorrosive coolants.

More detailed explanations can be found in the NEMA Standard 250 and in the UL White Book.

EXAMPLE:
Select an enclosure suitable for use indoors only.
Answer: NEMA 1 enclosure.

PRACTICE EXERCISE 1: Select an enclosure that could be used for normal outdoor applications.

PRACTICE EXERCISE 2: Select an enclosure for indoor use where there will be a limited amount of dust and dirt.

PRACTICE EXERCISE 3: Select an enclosure for use in a car wash application where water will spray against the enclosure at times.

1.6 GROUND RODS, MULTIPLE

The *National Electrical Code®, Section 250-84*, discusses the maximum permitted resistance to ground for driven ground rods. The following formula can be used to determine, with reasonable accuracy, the combined resistance to ground when driving multiple ground rods placed at least one rod length apart.

Number of Rods	Multiplying Factor (MF)
1	1.00
2	1.16
3	1.29
4	1.36
8	1.68
12	1.80
16	1.92
20	2.00
24	2.16

The formula for combined resistance to ground for multiple rods (R_t) is:

$$\mathbf{R_t = \frac{Resistance\ ground\ for\ one\ rod}{Number\ of\ rods} \times Multiplying\ factor\ from\ chart}$$

above

$$\mathbf{R_t = \frac{R_1}{N} \times MF}$$

EXAMPLE:

The resistance to ground for one driven ground rod is found to be 30 ohms. Assuming proper spacing and depth, and similar soil and moisture, determine the approximate resistance of two rods solidly connected to the same grounding electrode conductor.

$$R_t = \frac{R_1}{N} \times MF$$

$$= \frac{30}{2} \times 1.16$$

$$= 15 \times 1.16$$
$$= 17.4 \text{ OHMS}$$

Section 250-84 states that the rods shall be spaced no less than six feet apart. Some engineering manuals suggest that the spacing shall be the same as the length of the rod. For example, if eight-foot rods are used, space them eight feet apart.

PRACTICE EXERCISE: Calculate the combined resistance for an installation of three ground rods. Assume depth, moisture, etc. around all of the rods is the same. The resistance to ground for one rod was measured, and found to be 20 ohms.

1.7 MOTOR CODE LETTERS

Table 430-7(b) of the *National Electrical Code*® shows the locked-rotor kVA values for different Code Letters. Code Letters are found on the nameplates of motors. The Code Letters allow us to calculate the momentary inrush current that a motor will draw when started. The formulas are:

Three-Phase Motors

$$\text{Amperes} = \frac{\text{kVA} \times 1{,}000}{\text{Volts} \times 1.73}$$

Single-Phase Motors

$$I = \frac{\text{kVA} \times 1{,}000}{\text{Volts}}$$

Code Letter On Nameplate	kVA per Horsepower with Locked Rotor
A	0–3.14
B	3.15–3.54
C	3.55–3.99
D	4.0–4.49
E	4.5–4.99
F	5.0–5.59
G	5.6–6.29
H	6.3–7.09
J	7.1–7.99
K	8.0–8.99
L	9.0–9.99
M	10.0–11.19
N	11.2–12.49
P	12.5–13.99
R	14.0–15.99
S	16.0–17.99
T	18.0–19.99
U	20.0–22.39
V	22.4–& up

EXAMPLE:

Calculate the minimum and maximum starting current for a 15-horsepower, 240-volt, 3-phase, Code Letter "J" motor.

Minimum inrush current:

$$I = \frac{kVA \times 1{,}000}{E \times 1.73}$$

$$= \frac{15 \times 7.1 \times 1{,}000}{240 \times 1.73}$$

$$= 256.5 \text{ AMPERES}$$

Maximum inrush current:

$$= \frac{15 \times 7.99 \times 1{,}000}{240 \times 1.73}$$

$$= 288.7 \text{ AMPERES}$$

PRACTICE EXERCISE 1: Find the minimum and maximum starting current of a 10-horsepower, 480-volt, 3-phase, Code Letter "N" motor.

PRACTICE EXERCISE 2: Find the minimum and maximum starting current for a 10-horsepower, 480-volt, 3-phase, Code Letter "D" motor.

1.8 NEUTRAL CURRENT CALCULATIONS: Three-phase, Four-wire Wye Connected Systems

The following procedure shows neutral current calculations for resistive loads. Inductive, capacitive, and non-linear and linear loads (harmonics) require complex calculations beyond the scope of this "hands-on" book.

Refer to Chapters 3 and 4 for "Rule of Thumb" multipliers that can be used for these types of systems.

Section 220-22 and Note 8 to *Table 310-16* of the *NEC*® indicate that the neutral of a three-wire circuit supplied from a three-phase, four-wire system carries approximately the same current as the phase conductors.

When there are L-L or L-L-L connected loads in these types of circuits, subtract these current values from the phase currents before calculating the neutral current values. These L-L and L-L-L loads DO NOT involve the neutral conductor.

EXAMPLE 1:

Determine the neutral current in the diagram. Phase "A" is carrying 60 amperes. Phase "B" is carrying 100 amperes. All loads are connected L-N. There are no L-L loads.

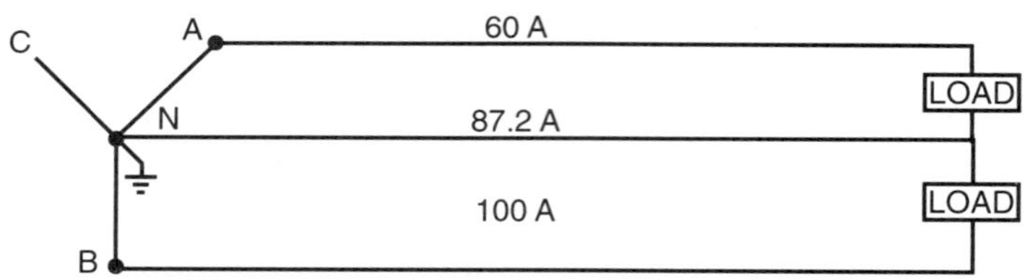

$$\begin{aligned} I_N &= \sqrt{(A^2 + B^2) - (AB)} \\ &= \sqrt{(60^2 + 100^2) - (60 \times 100)} \\ &= \sqrt{3{,}600 + 10{,}000 - 6{,}000} \\ &= \sqrt{7{,}600} \\ &= 87.2 \text{ AMPERES} \end{aligned}$$

PRACTICE EXERCISE 1: Determine the neutral current (I_N) in a 3-wire feeder supplied from a wye connected 3-phase, 4-wire system. The load is resistive. The current in Phase "A" is 16 amperes. The current in Phase "B" is 16 amperes. The loads are connected L-N. There are no L-L connected loads.

EXAMPLE 2:

Calculate the neutral current (I_N) on a 3-phase, 4-wire (3ϕ, 4W) wye connected system circuit. The loads are resistive. The phase currents are A = 10, B = 20, C = 30. The loads are connected L-N. There are no L-L or L-L-L connected loads.

$$
\begin{aligned}
I_N &= \sqrt{A^2 + B^2 + C^2) - AB - BC - CA} \\
&= \sqrt{(10^2 + 20^2 + 30^2) - (10 \times 20) - (20 \times 30) - (30 \times 10)} \\
&= \sqrt{100 + 400 + 900 - 200 - 600 - 300} \\
&= \sqrt{300} \\
&= 17.32 \text{ AMPERES}
\end{aligned}
$$

PRACTICE EXERCISE 2: Calculate the neutral current (I_N) on a 3ϕ, 4W wye connected feeder. The phase currents are A = 60, B = 80, C = 100. The loads are connected L-N. There are no L-L or L-L-L connected loads.

1.9 CENTIGRADE (CELSIUS)/FAHRENHEIT TEMPERATURE CONVERSIONS

The *National Electrical Code*® gives temperature ratings for conductors in both Celsius and Fahrenheit scales. The following shows some of the more common temperature ratings found in the Code.

CELSIUS	FAHRENHEIT	
0°C =	32°F	Water freezes at sea level.
30°C =	86°F	Normal ambient temperature for conductor ampacity (see *Table 310-16*, *NEC*®).
60°C =	140°F	Typical temperature ratings for conductor ampacity (see *Table 310-16*, *NEC*®).
75°C =	168°F	
90°C =	194°F	
100°C =	212°F	Water boils at sea level.

1.9.1 Converting Celsius to Fahrenheit

$$\mathbf{F = \left(\frac{9 \times C}{5}\right) + 32}$$

EXAMPLE:

Find F for 60°C.

$$\begin{aligned} F &= \left(\frac{9 \times 60}{5}\right) + 32 \\ &= \left(\frac{540}{5}\right) + 32 \\ &= 108 + 32 \\ &= 140^\circ\ C \end{aligned}$$

PRACTICE EXERCISE: Convert 100°C to Fahrenheit.

1.9.2 Converting Fahrenheit to Celsius

$$\mathbf{C = \frac{(F - 32) \times 5}{9}}$$

EXAMPLE:
Find C for 212°F.

$$C = \frac{(F - 32) \times 5}{9}$$
$$= \frac{(212 - 32) \times 5}{9}$$
$$= \frac{180 \times 5}{9}$$
$$= \frac{900}{9}$$
$$= 100^\circ \text{ C}$$

PRACTICE EXERCISE: Change 194°F to Celsius.

AVERAGE MELTING TIME-CURRENT CHARACTERISTIC CURVES
LPN-RK AND LPS-RK LOW-PEAK FUSES

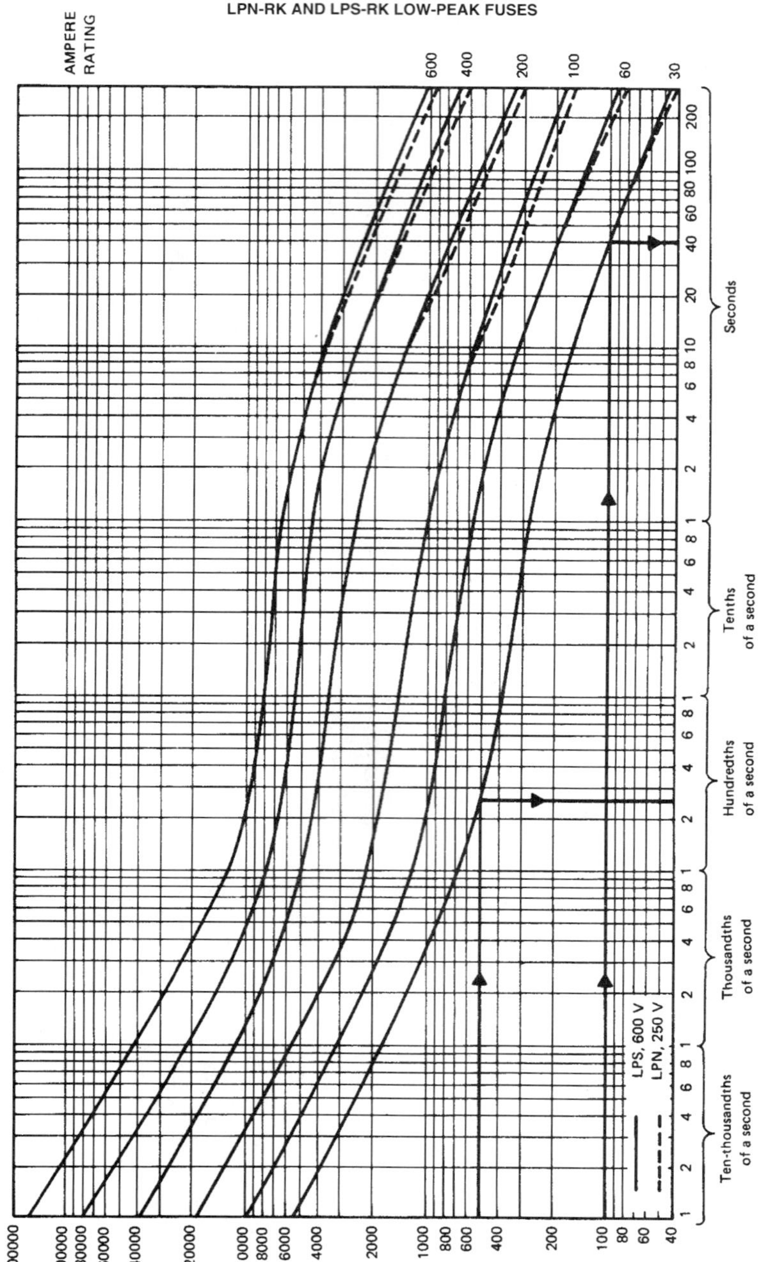

PROSPECTIVE CURRENT IN RMS SYMMETRICAL AMPERES

1.10 TIME/CURRENT CURVES

Time/current curves for fuses and circuit breakers are used to determine how long it will take the fuse or breaker to open the circuit when a certain amount of current is flowing.

The UL standard for time-delay fuses is that the fuse must "hold" five times its ampere rating for a *minimum* of ten seconds. For some low ampere rating fuses, the minimum time is eight seconds. Larger fuses will hold five times their rating for longer than ten seconds.

The time/current curve shown here is typical for a time-delay, dual-element fuse. This curve represents the average melting time for this type of time-delay, dual-element fuse. Other types of curves are available that show "minimum melting time" and/or "total clearing time."

Time/current curves for circuit breakers are similar in appearance.

EXAMPLE 1:
How long will a 30-ampere time-delay, dual-element fuse carry 100 amperes?
ANSWER: Enter the left side of the chart at 100 amperes. Follow along that line to the right until it intersects with the 30-ampere fuse line. Drop down to the base line, and find that the 30-ampere fuse will hold 100 amperes for approximately 40 seconds.

PRACTICE EXERCISE 1: How long will a 30-ampere time-delay, dual-element fuse hold 500 amperes?

PRACTICE EXERCISE 2: If a 600-ampere fuse of the type shown in the time/current curve is subjected to a 20,000-ampere short circuit, approximately how long will it take the fuse to clear the fault?

PRACTICE EXERCISE 3: How long will a 60-ampere time-delay, dual-element fuse hold a motor's inrush current of 300 amperes?

1.11 TRANSFORMER CALCULATIONS

The following examples show how to calculate current in amperes (I) when kVA is known and how to calculate kVA when current in amperes is known.

1.11.1 Transformers—Full Load Current Ratings Three-Phase, Dry Type Transformers

$$\text{kVA} = \frac{\text{Volts} \times \text{Amperes} \times 1.73}{1{,}000} \qquad \text{Amperes} = \frac{\text{kVA} \times 1{,}000}{\text{Volts} \times 1.73}$$

Kva Rating	Rated Line Voltage									
	120	208	240	480	600	2400	4160	7200	12470	13200
25	120	69.5	60	30	24	6	3.5	2	1.16	1.1
37.5	180	104	90	45	36	9	5.2	3	1.74	1.6
50	240	138	120	60	48	12	7	4	2.3	2.2
75	360	208	180	90	72	18	10.4	6	3.5	3.3
100	480	278	240	120	96	24	14	8	4.6	4.3
112.5	540	312	270	135	108	27	15.6	9	5.2	4.9
150	720	415	360	180	144	36	21	12	6.9	6.6
200	960	554	480	240	192	48	28	16	9.2	8.6
225	1080	625	540	270	216	54	31.2	18	10.4	9.8
250	1200	695	600	300	240	60	35	20	11.6	10.8
300	1440	830	720	360	288	72	42	24	13.9	13.2
400	1920	1110	960	480	384	96	55.6	32	18.5	17.5
500	2400	1380	1200	600	480	120	70	40	23.1	22
600	2880	1660	1440	720	576	144	84	48	27.7	26.2
750	3600	2080	1800	900	720	180	104	60	34.7	33
1000	4800	2780	2400	1200	960	240	140	80	46.2	44
1500	7200	4150	3600	1800	1440	360	208	120	69.4	66
2000	9600	5540	4800	2400	1920	480	278	160	92.4	87
2500	12042	6948	6021	3011	2409	602	347	201	116	110

1.11.2 Transformers—Full Load Current Ratings Single-Phase, Dry Type Transformers

$$\text{kVA} = \frac{\text{Volts} \times \text{Amperes}}{1{,}000} \qquad \text{Amperes} = \frac{\text{kVA} \times 1{,}000}{\text{Volts}}$$

Kva Rating	Rated Line Voltage								
	120	240	277	480	600	2400	4160	7200	13200
15	124	62	54	31	25	6.2	3.6	2.1	1.14
25	208	104	90	52	42	10.4	6	3.48	1.9
37.5	312	156	135	78	62	15.6	9	5.2	2.8
50	416	208	180	104	84	21	12	6.9	3.8
75	624	312	270	156	124	31	18	10.4	5.6
100	830	415	360	207	168	42	24	13.9	7.5
125	1040	520	450	260	208	52	30	17.3	9.5
150	1248	624	540	312	248	62	36	20.8	11.8
167	1390	695	601	348	278	70	40	23.2	12.6
200	1660	833	720	416	336	84	48	27.8	15.0
225	1875	938	812	469	375	94	54	31.3	17.0
250	2080	1040	900	520	420	105	60	34.8	19.0
300	2500	1250	1083	625	500	125	72	41.7	22.7
333	2780	1390	1199	695	555	139	80	46	25
400	3320	1660	1440	830	672	168	96	55.6	30
500	4160	2080	1800	1040	840	210	120	69.5	38
600	5000	2500	2160	1250	1000	250	144	83.6	45
750	6240	3120	2700	1560	1240	310	180	104	57
1000	8300	4150	3600	2075	1680	420	240	139	76

1.11.3 Single-Phase Transformer

EXAMPLE 1:

A 120-volt single-phase transformer is needed to supply power to a group of lights. The load demand will be approximately 83 amperes. What kVA rating should the transformer be?

$$\text{kVA} = \frac{\text{Volts} \times \text{Amperes}}{1{,}000}$$

$$= \frac{120 \times 83}{1{,}000}$$

$$= \frac{9{,}960}{1{,}000} = 9.96$$

This would require a minimum transformer rating of 10 kVA.

PRACTICE EXERCISE 1: Calculate the kVA line-to-neutral load on a 277-volt system that is carrying 200 amperes.

EXAMPLE 2:

A transformer is rated 167 kVA, 240/120 volts. Calculate the full-load current output that this transformer is capable of delivering at 240 volts.

$$I = \frac{kVA \times 1{,}000}{E}$$

$$= \frac{167 \times 1{,}000}{240}$$

$$= \frac{167{,}000}{240} = 696 \text{ AMPERES}$$

PRACTICE EXERCISE 2: Show the calculations for determining the full-load current output for a 500 kVA, 480-volt single-phase transformer.

PRACTICE EXERCISE 3: Show the calculations for determining the intended supply voltage for a 150 kVA transformer, single-phase, 312-ampere secondary.

1.11.4 Three-Phase Transformers

EXAMPLE 1:

Show the calculations for determining the kVA ratings of a 3-phase, 2,400-volt transformer marked 120 amperes.

$$kVA = \frac{\text{Volts} \times \text{Amperes} \times 1.73}{1{,}000}$$

$$= \frac{2400 \times 120 \times 1.73}{1{,}000}$$

$$= \frac{498{,}240}{1{,}000}$$

$$= 498.24$$

(This is a 500 kVA transformer.)

PRACTICE EXERCISE 1: A small 3-phase, 480-volt transformer must supply a 12-ampere load. Convert this information into kVA rating.

EXAMPLE 2:

Show the calculations for finding the current rating of a 750 kVA, 4160-volt 3-phase transformer.

$$I = \frac{kVA \times 1{,}000}{E \times 1.73}$$
$$= \frac{750 \times 1{,}000}{4{,}160 \times 1.73}$$
$$= \frac{750{,}000}{7{,}197}$$
$$= 104 \text{ AMPERES}$$

PRACTICE EXERCISE 2: Find the current rating of a 333 kVA, 480-volt, 3-phase transformer.

PRACTICE EXERCISE 3: What supply voltage is needed for a 112.5 kVA with a primary current of 312 ampere transformer? Show all calculations.

1.12 RIGHT TRIANGLES

The following information is needed to determine circuit Impedance (Z), when Resistance (R) and Reactance (X) are known.

Sines and cosines are discussed in this book in the section that covers alternating current circuit calculations. Pocket calculators provide an easy method for determining the values of sines, cosines, arc sines, etc.

The following right triangles show the more common ways to solve for the adjacent side, the opposite side, the hypotenuse, the sine, cosine, etc. when at least two factors are known.

The relationships of the hypotenuse, opposite, and adjacent are simple. Just remember three words (OAT, ACH, OSH). Basic math will then enable you to solve for one value when two other values are known.

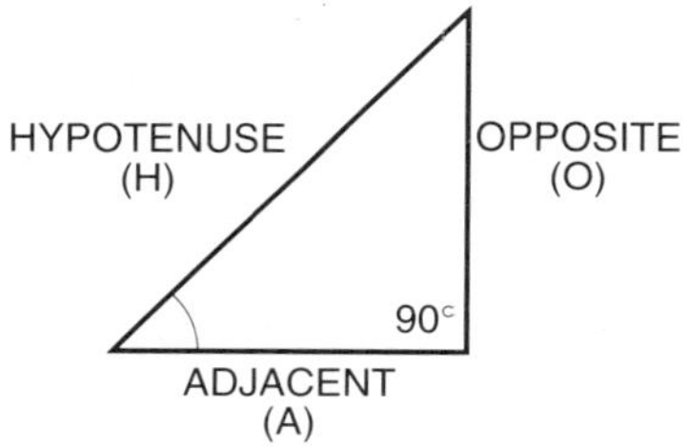

$$\mathbf{O = AT}$$

means Opposite = Adjacent × Tangent

then $$\mathbf{A = \frac{O}{T}}$$

$$\text{means Adjacent} = \frac{\text{Opposite}}{\text{Tangent}}$$

and $$\mathbf{T = \frac{O}{A}}$$

$$\text{means Tangent} = \frac{\text{Opposite}}{\text{Adjacent}}$$

$$\mathbf{A = CH}$$

means Adjacent = Cosine × Hypotenuse

then $\mathbf{C = \frac{A}{H}}$

means $\text{Cosine} = \frac{\text{Adjacent}}{\text{Hypotenuse}}$

and $\mathbf{H = \frac{A}{C}}$

means $\text{Hypotenuse} = \frac{\text{Adjacent}}{\text{Cosine}}$

$$\mathbf{O = SH}$$

means Opposite = Sine × Hypotenuse

then $\mathbf{S = \frac{O}{H}}$

means $\text{Sine} = \frac{\text{Opposite}}{\text{Hypotenuse}}$

and $\mathbf{H = \frac{O}{S}}$

means $\text{Hypotenuse} = \frac{\text{Opposite}}{\text{Sine}}$

How To Calculate Circuit Impedance.

$$\text{Circuit Impedance} = \sqrt{\text{Circuit Resistance}^2 \times \text{Circuit Reactance}^2}$$

$$Z = \sqrt{R^2 \times X^2}$$

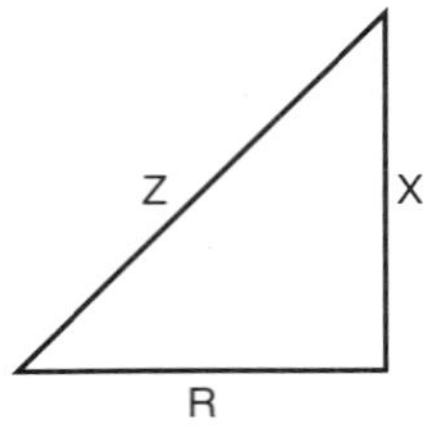

EXAMPLE:

Solve for the impedance (Z) of a circuit having a resistance (R) of 3 ohms and a reactance (X) of 4 ohms.

$$
\begin{aligned}
Z &= \sqrt{R^2 + X^2} \\
&= \sqrt{3^2 + 4^2} \\
&= \sqrt{9 + 16} \\
&= \sqrt{25} \\
&= 5 \text{ OHMS impedance}
\end{aligned}
$$

PRACTICE EXERCISE 1: A conductor has a resistance value of 0.12 ohms (R) and a reactance value of 0.003 ohms (X). Calculate the conductor's impedance (Z).

PRACTICE EXERCISE 2: Find the approximate length of the guy wire in the diagram.

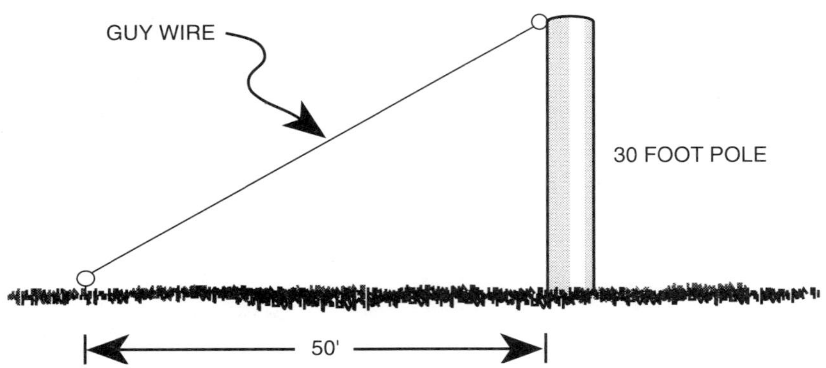

PRACTICE EXERCISE 3: A pull box will have two 3-inch conduits enter the box. The *NEC*® requires that the distance between raceways shall not be less than six times the trade diameter of the raceway.
Calculate for distance "X."

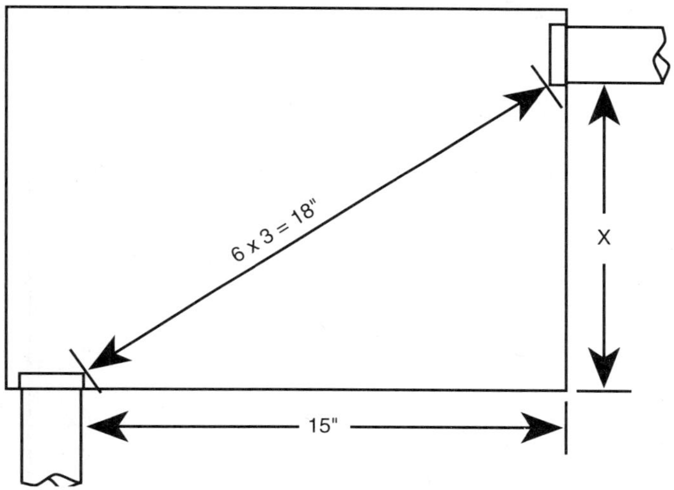

PRACTICE EXERCISE 4: The following floor plan indicates that a conduit must be run from Panel A to Panel B. If the conduit can be installed in the concrete, run at an angle, a considerable amount of conduit, conductors, and labor can be saved in completing the installation. If the conduit is installed after the concrete pour, the conduit must be run on the surface, and it must be run at right angles for appearance purposes. Calculate how much conduit would be saved by putting it in the concrete rather than surface mounting the conduit. A surface installation will require ten feet of conduit from the top of the panels to the ceiling.

The distance from the bottom of the panel to in or under the concrete is four feet. Don't concern yourself with the length of the elbows (or bends) required. Just determine the approximate conduit runs for both types of installations.

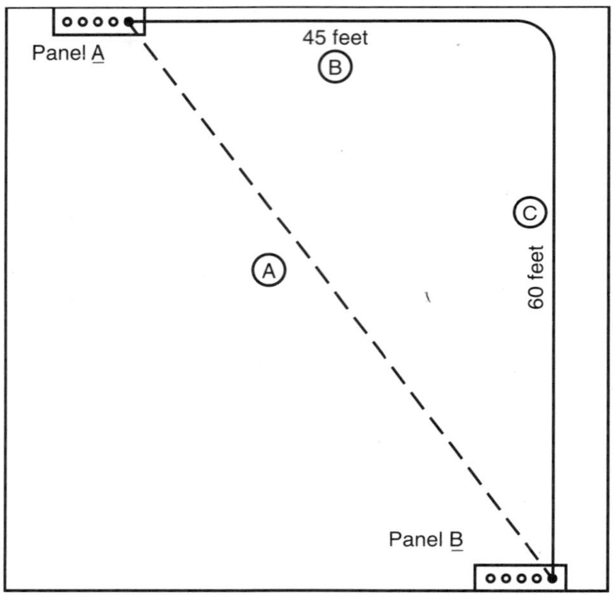

1.13 SHORT-CIRCUIT CURRENT CALCULATIONS AT TRANSFORMER SECONDARY TERMINALS

For simplicity, no motor contribution has been added in the following examples. Motors act as generators when a system has been short-circuited. They contribute short-circuit current into the system, which is added to the fault current delivered by the utility. Motors contribute short-circuit current approximately equal to their starting current. This can vary from four to eight times the motor's full-load running current. It greatly depends upon the Code Letter of motor. Refer to Unit 7 for calculations of motor starting (inrush) currents.

Also, for simplicity, this formula assumes that the utility can deliver unlimited fault current. This is sometimes referred to as an "infinite primary."

Step 1:

Determine the full-load rating of the transformer—in amperes—from the transformer nameplate, tables, or the following formula:

a) For three-phase transformers:

$$\mathbf{I_{FLA} = \frac{kVA \times 1{,}000}{E_{L-L} \times 1.73}}$$

where $E_{L\text{-}L}$ = line-to-line voltage

EXAMPLE 1: A 3ϕ, 208-volt, 167 kVA transformer.

$$I_{FLA} = \frac{167 \times 1{,}000}{208 \times 1.73} = 464 \text{ AMPERES}$$

b) For single-phase transformers:

$$\mathbf{I_{FLA} = \frac{kVA \times 1{,}000}{E_{L-L}}}$$

where $E_{L\text{-}L}$ = line-to-line voltage

EXAMPLE 2: A single-phase, 208-volt, 167 kVA transformer.

$$I_{FLA} = \frac{167 \times 1{,}000}{208} = 696 \text{ AMPERES}$$

Step 2:

Check nameplate on transformer for impedance (Z) value. A transformer nameplate is marked 2% impedance (Z).

Step 3:

Calculate transformer multiplier "M_1:"

$$\mathbf{M_1 = \frac{100}{Transformer\ Z \times 0.9}}$$

EXAMPLE 3: $\mathbf{M_1 = \frac{100}{2 \times 0.9} = 55.6}$

(Note: Because test standards for transformers allow ± 10% variance of *actual* to *marked* impedance, use 0.9 to reflect worst case conditions.)

Step 4:

Calculate short-circuit amperes at transformer secondary. Use the 167 kVA, 3ϕ, 208-Volt transformer current rating.

EXAMPLE 4: $\mathbf{I_{SCA} = transformer\ full\ load\ amperes \times multiplier}$

$I_{SCA} = 464 \times 55.6$

$= 25{,}798$ amperes of short-circuit current available at transformer secondary terminals

PRACTICE EXERCISE: Calculate the available short-circuit current at the secondary terminals of a 500 kVA, 3ϕ, 480-volt transformer. The transformer is marked 1.8% impedance.

1.14 SHORT-CIRCUIT CURRENT CALCULATIONS POINT-TO-POINT BEYOND TRANSFORMER

A simple method of determining the available short-circuit current (also referred to as fault current) at various distances from a given location is the "POINT-TO-POINT" method. Reasonable accuracy is obtained when this method is used with three-phase and single-phase systems.

The following procedure demonstrates the use of the "POINT-TO-POINT" method. "C" tables and "M" chart follow Step 7.

Step 1:

Determine the full-load rating of the transformer—in amperes—from the transformer nameplate, tables, or the following formulas:

a) For three-phase transformers:

$$I_{FLA} = \frac{\text{kVA} \times 1000}{E_{L-L} \times 1.73}$$

where $E_{L\text{-}L}$ = line-to-line voltage

b) For single-phase transformers:

$$I_{FLA} = \frac{\text{kVA} \times 1000}{E_{L-L}}$$

Step 2:

Find the percent impedance (Z) on nameplate of transformer.

Step 3:

Find the transformer multiplier "M_1:"

$$M_1 = \frac{100}{\text{Transformer } Z \times 0.9}$$

(Note: Because the marked transformer impedance can vary ± 10% per the UL standard, the 0.9 factor is used to show "worst case" conditions.)

Step 4:

Determine the transformer let-through short-circuit current at the secondary terminals of transformer. Use tables or the following formula:

a) For three-phase transformers (L-L-L):

$$I_{SCA_{L-L-L}} = \text{transformer}_{FLA} \times \text{multiplier "}M_1\text{"}$$

b) For single-phase transformers (L-L):

$$\mathbf{I_{SCA_{L-L}} = transformer_{FLA} \times multiplier \text{ ``}M_1\text{''}}$$

c) For single-phase transformers (L-N):

$$\mathbf{I_{SCA_{L-N}} = transformer_{FLA} \times multiplier \text{``}M_1\text{''} \times 1.5}$$

(Note: The 1.5 factor is explained in Step 5 in the paragraph marked "X.")

Step 5:

Determine the "f" factor:

a) For three-phase faults:

$$\mathbf{f = \frac{1.73 \times L \times I_{SCA_{L-L-L}}}{N \times C \times E_{L-L}}}$$

b) For single-phase, line-to-line (L-L) faults on single-phase, center-tapped transformers:

$$\mathbf{f = \frac{2 \times L \times I_{SCA_{L-L}}}{N \times C \times E_{L-L}}}$$

c) For single-phase, line-to-neutral (L-N) faults on single-phase, center-tapped transformers:

$$\mathbf{f = \frac{2 \times L \times I_{SCA_{L-N}}}{N \times C \times E_{L-N}}}$$

where

L = the length of the circuit to the fault, in feet.

I = the available fault current—in amperes—at the beginning of the circuit.

C = the constant derived from the tables for the specific type of conductors and wiring method.

E = the voltage, line-to-line or line-to-neutral. (See Step 4a, b, and c to decide which voltage to use.)

N = the number of conductors in parallel.

terminals of a single-phase, center-tapped transformer. At some distance from the terminals, depending upon the wire size, the L-N fault current is lower than the L-L fault current. This can vary from 1.33 to 1.67 times. These figures are based on the change in the turns ratio between the primary and secondary infinite source impedance, a distance of zero feet from the terminals of the transformer, and 1.2% reactance (X) and 1.5% resistance (R) for the L-N vs. L-L resistance and reactance values. For simplicity, in Step 4c we used an approximate multiplier of 1.5. First, do the L-N calculation at the transformer secondary terminals, Step 4c, then proceed with the point-to-point method.

Step 6:

After finding the "f" factor, refer to Chart M and locate the appropriate value of the multiplier "M_2." for the specific "f" value. Or, calculate as follows:

$$\mathbf{M_2 = \frac{1}{1+f}}$$

Step 7:

Multiply the available fault current at the beginning of the circuit by the multiplier "M_2" to determine the available symmetrical fault current at the fault.

$$\mathbf{I_{SCA}\text{ at fault} = I_{SCA}\text{ at beginning of circuit} \times \text{“}M_2\text{”}}$$

Note 1:

To make calculations to other points in the system, repeat Steps 5, 6, and 7.

Note 2:

Motor contribution may be added to the transformer secondary short-ciruit as calculated in Step 4. Typically, use 4X motor FLA for standard motors . . . . . . add 5X motor FLA for energy efficient motors. These values result in a reasonable approximation of motor contribution to the overall short circuit current calculations.

"C" VALUES

Copper Conductors

AWG or kcmil	Three Single Conductors: Steel Conduit 600V	Three Single Conductors: Steel Conduit 5KV	Three Single Conductors: Steel Conduit 15KV	Three Single Conductors: Nonmagnetic Conduit 600V	Three Single Conductors: Nonmagnetic Conduit 5KV	Three Single Conductors: Nonmagnetic Conduit 15KV	Three Conductor Cable: Steel Conduit 600V	Three Conductor Cable: Steel Conduit 5KV	Three Conductor Cable: Steel Conduit 15KV	Three Conductor Cable: Nonmagnetic Conduit 600V	Three Conductor Cable: Nonmagnetic Conduit 5KV	Three Conductor Cable: Nonmagnetic Conduit 15KV
14	389	—	—	389	—	—	389	—	—	389	—	—
12	617	—	—	617	—	—	617	—	—	617	—	—
10	981	—	—	981	—	—	981	—	—	981	—	—
8	1557	1551	1557	1556	1555	1558	1559	1557	1559	1559	1558	1559
6	2425	2406	2389	2430	2417	2406	2431	2424	2414	2433	2428	2420
4	3806	3750	3695	3825	3789	3752	3830	3811	3778	3837	3823	3798
3	4760	4760	4760	4802	4802	4802	4760	4790	4760	4802	4802	4802
2	5906	5736	5574	6044	5926	5809	5989	5929	5827	6087	6022	5957
1	7292	7029	6758	7493	7306	7108	7454	7364	7188	7579	7507	7364
1/0	8924	8543	7973	9317	9033	8590	9209	9086	8707	9472	9372	9052
2/0	10755	10061	9389	11423	10877	10318	11244	11045	10500	11703	11528	11052
3/0	12843	11804	11021	13923	13048	12360	13656	13333	12613	14410	14118	13461
4/0	15082	13605	12542	16673	15351	14347	16391	15890	14813	17482	17019	16012
250	16483	14924	13643	18593	17120	15865	18310	17850	16465	19779	19352	18001
300	18176	16292	14768	20867	18975	17408	19557	20051	18318	22524	21938	20163
350	19703	17385	15678	22736	20526	18672	20617	21914	19821	22736	24126	21982
400	20565	18235	16365	24296	21786	19731	24253	23371	21042	26915	26044	23517
500	22185	19172	17492	26706	23277	21329	26980	25449	23125	30028	28712	25916
600	22965	20567	17962	28033	25203	22097	28752	27974	24896	32236	31258	27766
750	24136	21386	18888	28303	25430	22690	31050	30024	26932	32404	31338	28303
1000	25278	22539	19923	31490	28083	24887	33864	32688	29320	37197	35748	31959

"C" VALUES

Aluminum Conductors

	Three Single Conductors						Three Conductor Cable					
AWG or kcmil	Steel Conduit			Nonmagnetic Conduit			Steel Conduit			Nonmagnetic Conduit		
	600V	5KV	15KV	600V	5KV	15KV	600V	5KV	15KV	600V	5KV	15KV
14	236	—	—	236	—	—	236	—	—	236	—	—
12	375	—	—	375	—	—	375	—	—	375	—	—
10	598	—	—	598	—	—	598	—	—	598	—	—
8	951	950	951	951	950	951	951	951	951	951	951	951
6	1480	1476	1472	1481	1478	1476	1481	1480	1478	1482	1481	1479
4	2345	2332	2319	2350	2341	2333	2351	2347	2339	2353	2349	2344
3	2948	2948	2948	2958	2958	2958	2948	2956	2948	2958	2958	2958
2	3713	3669	3626	3729	3701	3672	3733	3719	3693	3739	3724	3709
1	4645	4574	4497	4678	4631	4580	4686	4663	4617	4699	4681	4646
1/0	5777	5669	5493	5838	5766	5645	5852	5820	5717	5875	5851	5771
2/0	7186	6968	6733	7301	7152	6986	7327	7271	7109	7372	7328	7201
3/0	8826	8466	8163	9110	8851	8627	9077	8980	8750	9242	9164	8977
4/0	10740	10167	9700	11174	10749	10386	11184	11021	10642	11408	11277	10968
250	12122	11460	10848	12862	12343	11847	12796	12636	12115	13236	13105	12661
300	13909	13009	12192	14922	14182	13491	14916	14698	13973	15494	15299	14658
350	15484	14280	13288	16812	15857	14954	15413	15490	15540	16812	17351	16500
400	16670	15355	14188	18505	17321	16233	18461	18063	16921	19587	19243	18154
500	18755	16827	15657	21390	19503	18314	21394	20606	19314	22987	22381	20978
600	20093	18427	16484	23451	21718	19635	23633	23195	21348	25750	25243	23294
750	21766	19685	17686	23491	21769	19976	26431	25789	23750	25682	25141	23491
1000	23477	21235	19005	28778	26109	23482	29864	29049	26608	32938	31919	29135

"C" Values

Ampacity	Plug-In Busway		Feeder Busway		High Imped. Busway
	Copper	Aluminum	Copper	Aluminum	Copper
225	28700	23000	18700	12000	—
400	38900	34700	23900	21300	—
600	41000	38300	36500	31300	—
800	46100	57500	49300	44100	—
1000	69400	89300	62900	56200	15600
1200	94300	97100	76900	69900	16100
1350	119000	104200	90100	84000	17500
1600	129900	120500	101000	90900	19200
2000	142900	135100	134200	125000	20400
2500	143800	156300	180500	166700	21700
3000	144900	175400	204100	188700	23800
4000	—	—	277800	256400	—

CHART M (multiplier)

f	M	f	M
0.01	0.99	1.20	0.45
0.02	0.98	1.50	0.40
0.03	0.97	2.00	0.33
0.04	0.96	3.00	0.25
0.05	0.95	4.00	0.20
0.06	0.94	5.00	0.17
0.07	0.93	6.00	0.14
0.08	0.93	7.00	0.13
0.09	0.92	8.00	0.11
0.10	0.91	9.00	0.10
0.15	0.87	10.00	0.09
0.20	0.83	15.00	0.06
0.30	0.77	20.00	0.05
0.40	0.71	30.00	0.03
0.50	0.67	40.00	0.02
0.60	0.63	50.00	0.02
0.70	0.59	60.00	0.02
0.80	0.55	70.00	0.01
0.90	0.53	80.00	0.01
1.00	0.50	90.00	0.01
		100.00	0.01

$$M = \frac{1}{1 + f}$$

EXAMPLE:

Calculate the available L-L-L short-circuit current (fault current) at X_1 X_2 in the one-line diagram.

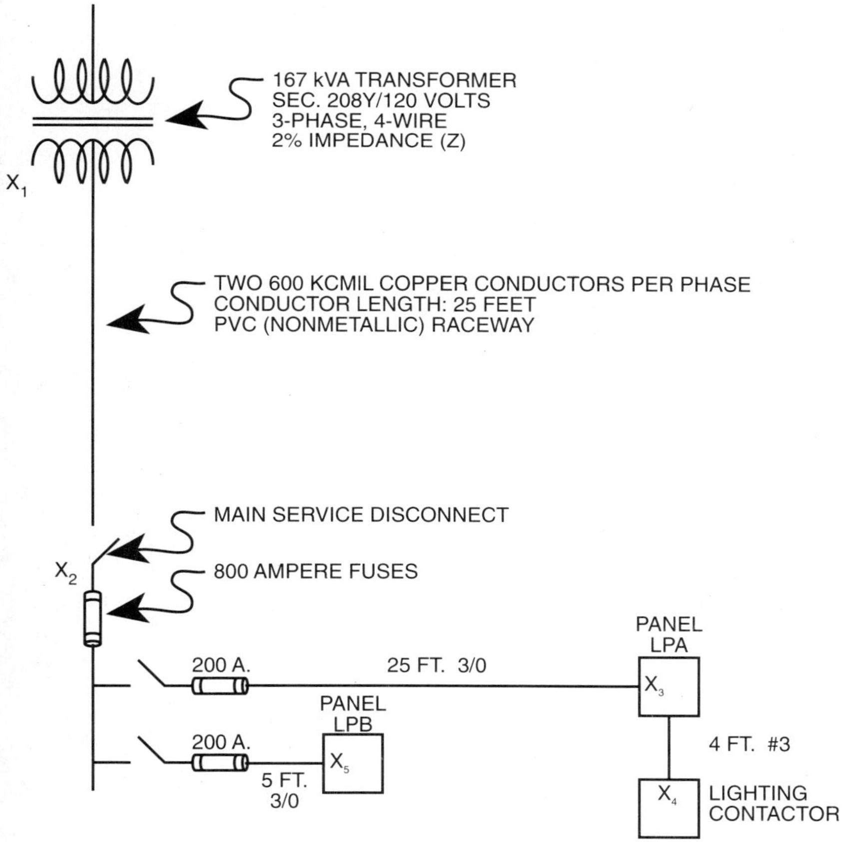

NOTE: SERVICE RACEWAY-PVC
ALL OTHER RACEWAYS-STEEL CONDUIT
ALL CONDUCTORS-COPPER

EXAMPLE CALCULATIONS:

Step 1:

Calculate full-load current of transformer.

$$I = \frac{kVA \times 1{,}000}{E \times 1.73} = \frac{167 \times 1{,}000}{208 \times 1.73} = \frac{167{,}000}{360} = 464 \text{ AMPERES}$$

Step 2:

Find the transformer impedance. It is 2% in this example.

Step 3:

Calculate transformer multiplier "M_1."

$$M_1 = \frac{100}{Z \times 0.9} = \frac{100}{2 \times 0.9} = 55.6$$

Step 4:

Calculate short-circuit current at secondary of transformer. This is "X_1" in diagram.

$$\begin{aligned} I_{SCA} &= \text{Transformer}_{FLA} \times M_1 \\ &= 464 \times 55.6 \\ &= 25{,}798 \text{ AMPERES} \end{aligned}$$

Step 5:

Determine "f" factor.

$$\begin{aligned} f &= \frac{1.73 \times L \times I_{SCA_{L-L-L}}}{N \times C \times E_{L-L}} \\ &= \frac{1.73 \times 25 \times 25{,}798}{2 \times 28{,}033 \times 208} \\ &= 0.0956773 \end{aligned}$$

Step 6:

Calculate "M_2."

$$M_2 = \frac{1}{1+f} = \frac{1}{1+0.0956773} = 0.9126774 \text{ (use 0.913)}$$

Step 7:

Calculate short-circuit current at end of the 600 kcmil conductors in main switch. This is X_2 in diagram.

$$I_{SCA} = I_{SCA} \text{ at beginning of conductors} \times \text{“}M_2\text{”}$$
$$= 25{,}798 \times 0.913$$
$$= 23{,}554 \text{ AMPERES}$$

PRACTICE EXERCISE: Calculate the available short-circuit current (L-L-L) at Panel LPA (X_3), Lighting Contactor (X_4), and Panel LPB (X_5).

CHAPTER 2

Alternating Current Formulas

2.1 VALUES OF ALTERNATING CURRENT AND VOLTAGES

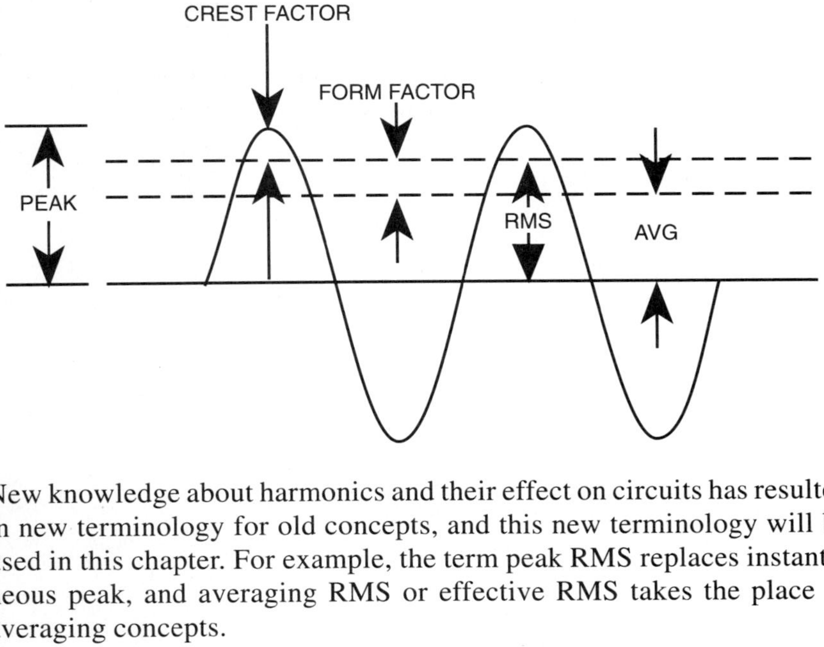

New knowledge about harmonics and their effect on circuits has resulted in new terminology for old concepts, and this new terminology will be used in this chapter. For example, the term peak RMS replaces instantaneous peak, and averaging RMS or effective RMS takes the place of averaging concepts.

$$\textbf{Peak RMS current or voltage} = \frac{\textbf{RMS}}{\textbf{0.707}}$$

$$\textbf{Effective RMS current or voltage} = I_{PEAK}\ (0.707) \text{ or } V_{PEAK}\ (0.707)$$

$$\textbf{Average RMS current or voltage} = I_{PEAK}\ (0.637) \text{ or } V_{PEAK}\ (0.637)$$

$$\textbf{Form Factor} = \frac{\textbf{RMS}}{\textbf{AVG}} = 1.11$$

$$\textbf{Crest Factor} = \frac{\textbf{PEAK}}{\textbf{RMS}} = 1.42$$

Full wave AC-DC rectification for voltage or current = (Peak) (0.637)

Half wave AC-DC rectification for voltage or

$$\textbf{current} = \frac{\textbf{(Peak)(0.637)}}{\textbf{2}}$$

EXAMPLE:

What is the no-load output DC voltage with a 120 V RMS input and full wave rectification?

$$\frac{120\text{ V}}{0.707} = (169.7\text{ V})(0.637) = 108\text{ VDC}$$

PRACTICE EXERCISE: What is the no-load output DC voltage with a 277 V RMS input and half wave rectification?

2.2 INDUCTANCE

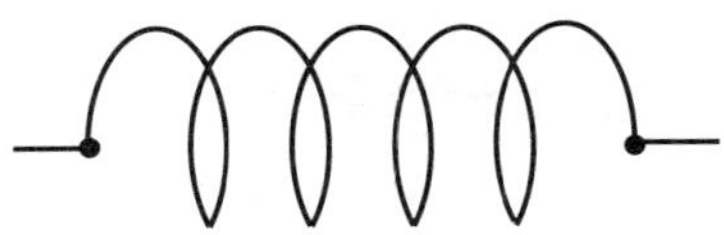

An inductance of one Henry (h) exists when current changing at the constant rate of one ampere per second results in a counter-electromotive-force (cemf) of one volt and its symbol is L.

Series: $\mathbf{L_T = L_1 + L_2 + L_3 + L_4 + \ldots}$

EXAMPLE 1:
3 coils where $L_1 = L_2 = L_3 = 5h$

$$L_T = 5h + 5h + 5h = 15h$$

Parallel: $\mathbf{\frac{1}{L_T} = \frac{1}{L_1} + \frac{1}{L_2} + \frac{1}{L_3} + \frac{1}{L_4} + \ldots}$

EXAMPLE 2:
3 coils of 5h each:

$$\frac{1}{L_T} = \frac{1}{5h} + \frac{1}{5h} + \frac{1}{5h} = \frac{3}{5h}$$

$$L = \frac{1}{\frac{3}{5h}} = 1 \div \frac{3}{5h} = 1 \times \frac{5h}{3} = 1.67h$$

PRACTICE EXERCISE: Find the series and parallel inductance of 4 coils having 3h, 4h, 5h, and 6h each.

2.3 CAPACITANCE

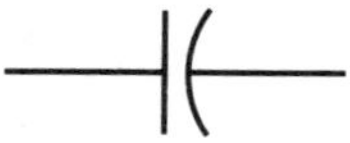

$$C = \frac{\textbf{Quantity of Charge in Coulombs(Q)}}{\textbf{Electromotive Force (E)}}$$

IMPORTANT RULE: 1 Coulomb = 6.28×10^{18} electrons

A capacitance (C) of one farad (f) exists when one volt of pressure is applied to plates causing one coulomb of charge to exist on those plates.

$$\textbf{Series: } \frac{1}{C_T} = \frac{1}{C_1} + \frac{1}{C_2} + \frac{1}{C_3} + \frac{1}{C_4} + \ldots$$

EXAMPLE 1:

Consider three capacitors, 6μf each:

$$\frac{1}{C_T} = \frac{1}{6\mu f} + \frac{1}{6\mu f} + \frac{1}{6\mu f} = \frac{3}{6\mu f} = \frac{1}{2\mu f}$$

$$C_T = \frac{1}{\frac{1}{2\ \mu f}} = 1 \div \frac{1}{2\mu f} = 1 \times \frac{2\mu f}{1} = 2\mu f$$

$$\textbf{Parallel: } C_T = C_1 + C_2 + C_3 + C_4 + \ldots$$

EXAMPLE 2:

Consider three capacitors, 6μf each:

$$C_T = 6\mu f + 6\mu f + 6\mu f = 18\mu f$$

PRACTICE EXERCISE: Find the total capacitance of 4 units in both series and parallel circuits rated 10μf, 20μf, 40μf, and 80μf.

2.4 CAPACITOR CHARGE/ DISCHARGE TIME CONSTANT

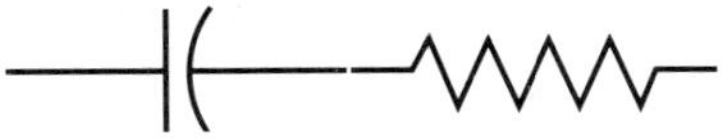

$$t = RC$$

where t = discharge /charge time

R = discharge/charge resistor

C = capacitance in μf

EXAMPLE:

What value resistor is required for a 480 V 1,000μf capacitor so it will comply with *Section 460-6* of the *National Electrical Code®*?

$$R = \frac{t}{C}$$

$$= \frac{15 \text{ seconds}}{0.001000 \text{ F}}$$

$$= 15{,}000 \text{ ohms}$$

PRACTICE EXERCISE: What value resistor is needed if the capacitor is 750μf and 250 V rated?

2.5 CAPACITOR CHARGE/DISCHARGE CURRENT

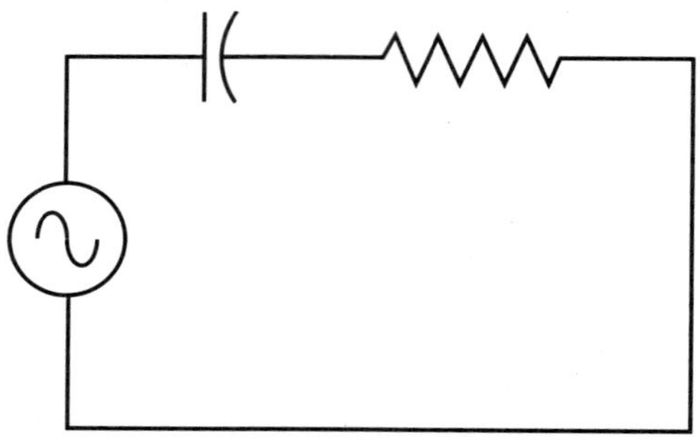

$$\mathbf{i} = \left(\frac{\mathbf{V}}{\mathbf{R}}\right) \times \left(\frac{\mathbf{1}}{\mathbf{2.718}^{\,t}}\right)$$

where t = time constantseach minute; no more than 15 seconds each
i = instantaneous current flow
V = voltage across the capacitor
R = value of resistor
2.718 = Natural log e

EXAMPLE 1:

What will be the current flow when shorting a 480 V 1,000µf capacitor with a screwdriver shank? Assume 0.1 ohm resistance and t = 0.

$$i = \left(\frac{480\text{ V}}{0.1\Omega}\right) \times \frac{1}{2.718^{\circ}}$$

$$= \left(\frac{480\text{ V}}{0.1\Omega}\right) \times \frac{1}{1}$$

$$= 4{,}800\text{ A}$$

EXAMPLE 2:

What will be the current flow when shorting a 480 V 1,000μf capacitor using a 15 k-ohm resistor at t = 0 and t = 1?

$$
\begin{aligned}
i &= \frac{480\ \text{V}}{15{,}000\ \Omega} \times \frac{1}{2.718^{\circ}} \\
&= \frac{480\ \text{V}}{15{,}000\ \Omega} \times \frac{1}{1} \\
&= 0.032\ \text{A}\ (32\ \text{mA}) \\
i &= \frac{480\ \text{V}}{15{,}000\ \Omega} \times \frac{1}{2.718^{1}} \\
&= \frac{480\ \text{V}}{15{,}000\,\Omega} \times \frac{1}{2.718} \\
&= 0.0117\ \text{A}\ (11.7\ \text{mA})
\end{aligned}
$$

PRACTICE EXERCISE: What will be the current flow using a 5 k-ohm resistor when t = 0 and t = 1?

2.6 CAPACITOR CHARGE/DISCHARGE VOLTAGE

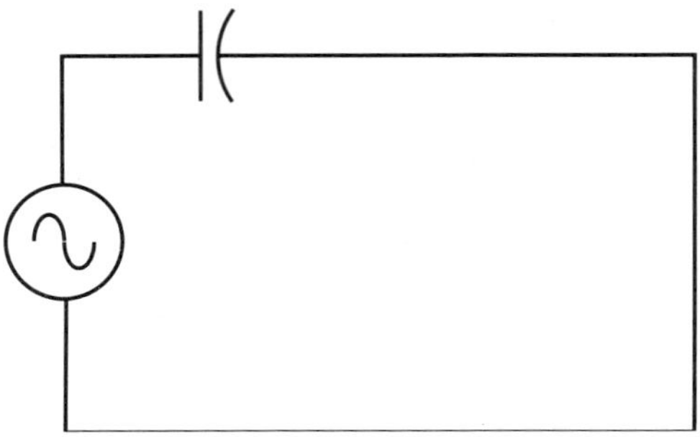

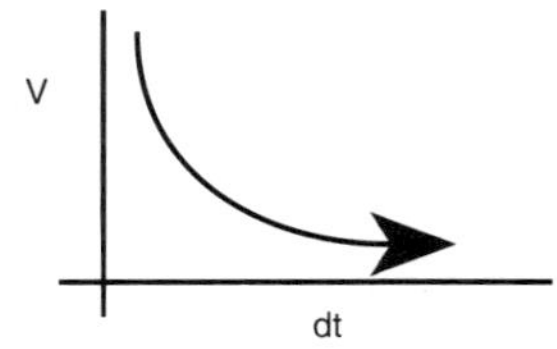

$$v_c = V_c \times \frac{1}{2.718^t}$$

where v_c = voltage across capacitor at t time constants

V_c = voltage when fully charged

t = number of time constants each minute; no more than 15 seconds each

EXAMPLE:

What will be the remaining voltage across a 480 V 1,000μf capacitor after the first minute?

$$v_c = 480\ V \times \frac{1}{2.718^{\frac{60}{15}}}$$

$$= 480\ V \times \frac{1}{2.718^4}$$

$$= 8.79\ V_c\ (8.8\ V)$$

PRACTICE EXERCISE: What will be the voltage after two minutes?

2.7 CAPACITOR CHARGE/DISCHARGE TIME CONSTANT/VOLTAGE

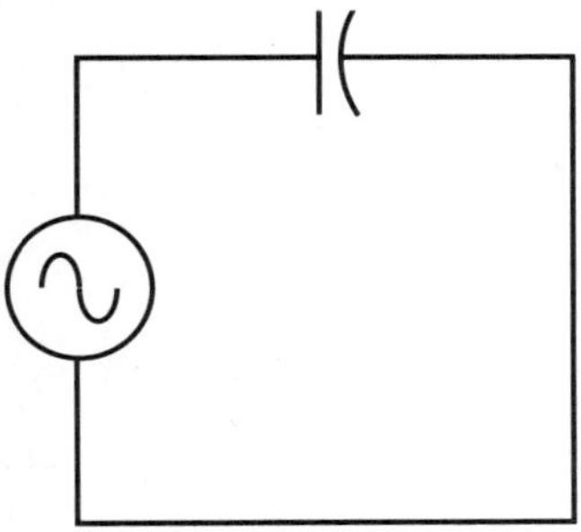

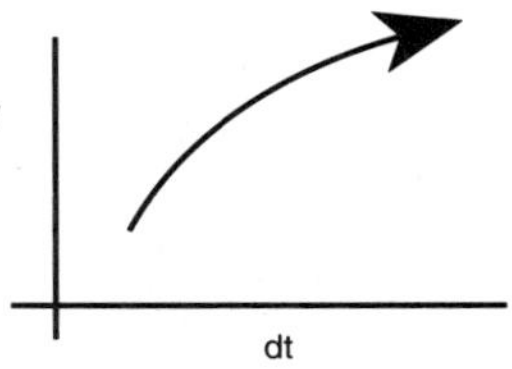

$$\mathbf{v_c = V_c\,[1 - (1 \div 2.718^{t})]}$$

where v_c = voltage across capacitor after t constants
V_c = circuit voltage
t = number of time constants each minute; no more than 15 seconds each

EXAMPLE:

What will be the voltage across a 480 V 1,000µf capacitor after the first minute of charging time?

$$\begin{aligned} v_c &= 480\ V_c\,[1 - (1 \div 2.718^{\frac{60}{15}})] \\ &= 480\ V_c\,[1 - (1 \div 2.718^{4})] \\ &= 471.2\ V_c\ (471\ V) \end{aligned}$$

PRACTICE EXERCISE: What is the voltage across a 250 volt 500µf capacitor after the first half-second?

2.8 REACTANCE

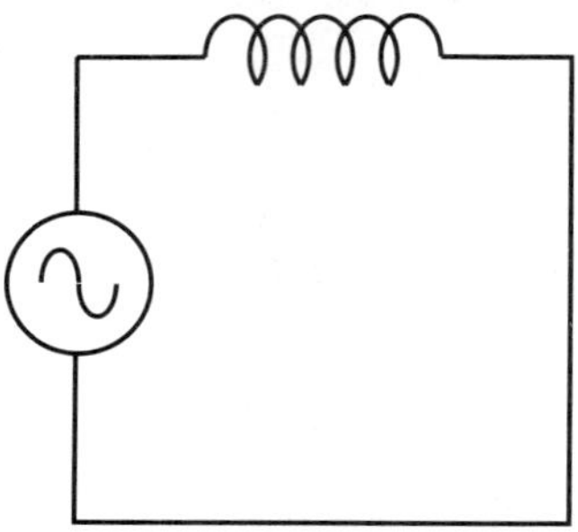

(a)

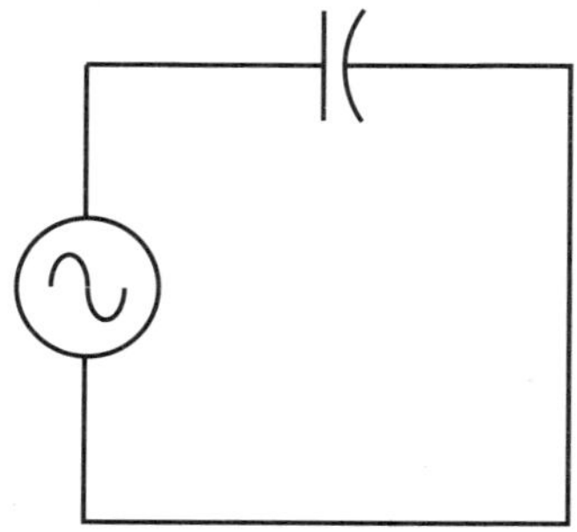

(b)

2.8.1 Inductive

$$\mathbf{X_L = 2\,\pi f L = \omega L}$$

where $\omega = 377$ at 60 Hz

$\omega = 157$ at 25 Hz

$\omega = 3{,}392$ at 540 Hz (9th harmonic of 60 Hz fundamental)

EXAMPLE:
The inductive reactance of a 5h coil at 60, 25, and 540 Hz will be:

$$X_L = 377 \times 5\text{h} = 1{,}885 \text{ ohms}$$
$$= 157 \times 5\text{h} = 785 \text{ ohms}$$
$$= 3{,}392 \times 5\text{h} = 16{,}960 \text{ ohms}$$

PRACTICE EXERCISE: Find the inductive reactance at 900 Hz and 5h (15th harmonic of 60 Hz).

2.8.2 Capacitive

$$\mathbf{Xc = \frac{1}{2\pi f C} = \frac{1}{\omega C}}$$

where $\omega = 377$ at 60 Hz

$\omega = 157$ at 25 Hz

$\omega = 3{,}392$ at 540 Hz (9th harmonic of 60 Hz fundamental)

EXAMPLE:

The capacitive reactance of a 6μf capacitor at 60, 25, and 540 Hz will be:

$$Xc = \frac{1}{377 \times 0.000006}$$
$$= \frac{1}{0.000262}$$
$$= 442 \text{ ohms}$$
$$= \frac{1}{157 \times 0.000006}$$
$$= \frac{1}{0.000262}$$
$$= 1{,}061 \text{ ohms}$$
$$= \frac{1}{3{,}392 \times 0.000006}$$
$$= \frac{1}{0.020352}$$
$$= 49 \text{ ohms}$$

PRACTICE EXERCISE: Find the capacitive reactance at 900 Hz and 6μf (15th harmonic of 60 Hz).

2.9 FREQUENCY

$$\mathbf{F = \frac{NP}{120} \qquad N = \frac{120F}{P} \qquad P = \frac{120F}{N}}$$

where F = frequency in Hertz
N = speed in RPM
P = number of poles
120 = constant based on 60 cycles per second and 2 alternations each second

EXAMPLE:
How many poles are there in a 1,200 RPM motor at 60 Hz?

$$P = \frac{120F}{N}$$

$$= \frac{(120)(60)}{1{,}200}$$

$$= 6$$

PRACTICE EXERCISE: What will be the RPM at 50 Hz?

2.10 POLYPHASE CIRCUITS

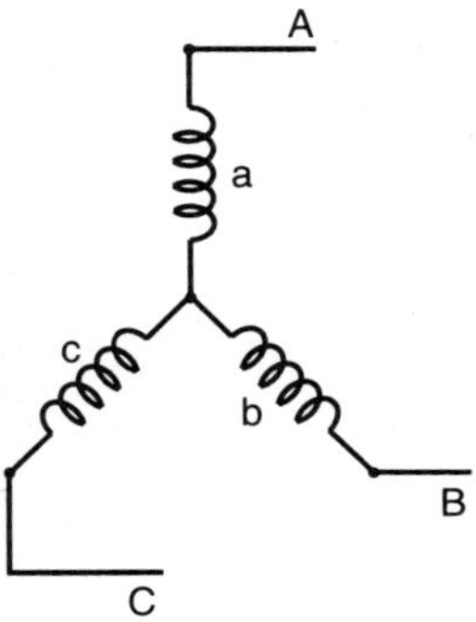

WYE (Y)
(a)

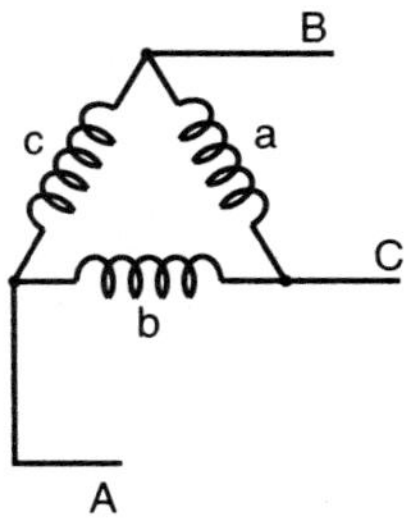

DELTA (Δ)
(b)

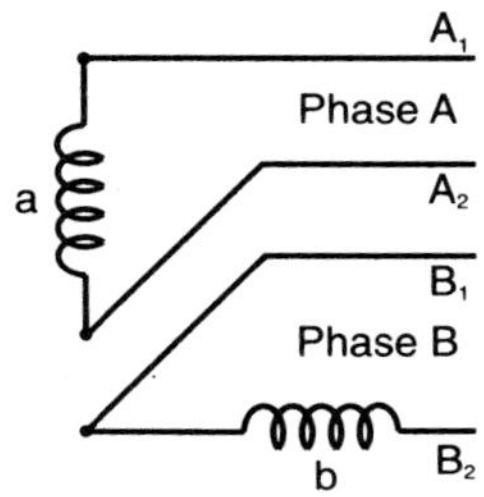

TWO-PHASE
(c)

2.10.1 Three-phase

Wye: $\mathbf{E_{L\text{-}L} = Ecoil \times 1.73}$

$\mathbf{I_A = Icoil_a}$

$\mathbf{I_B = Icoil_b}$

$\mathbf{I_C = Icoil_c}$

Delta: $\mathbf{E_{L\text{-}L} = Ecoil}$

$\mathbf{I_A = (Icoil_a + Icoil_b)(0.577)}$

$\mathbf{I_B = (Icoil_b + Icoil_c)(0.577)}$

$\mathbf{I_C = (Icoil_a + Icoil_c)(0.577)}$

2.10.2 Two-Phase

Four-Wire: $\mathbf{E_A = Ecoil_A}$

$\mathbf{E_B = Ecoil_B}$

$\mathbf{I_{A_1} = I_{A_2} = Icoil_A}$

$\mathbf{I_{B_1} = I_{B_2} = Icoil_B}$

Three-Wire: **(A_2 and B_1 connected)**

$$\mathbf{E_{A_1} = E_{A_2 B_1} = Ecoil_A\left(\frac{2}{\sqrt{3}}\right)}$$

$$\mathbf{E_{B_2} = E_{A_2 B_1} = Ecoil_B\left(\frac{2}{\sqrt{3}}\right)}$$

$$\mathbf{Icommon = I_{A_1} \times 1.414 = I_{A_2 B_1}}$$

or

$$\mathbf{= I_{B_2} \times 1.414 = I_{A_2 B_1}}$$

2.11 TRANSFORMER RATIOS

$$\frac{Np}{Ns} = \frac{Is}{Ip} \qquad \frac{Np}{Ns} = \frac{Ep}{Es} \qquad \frac{Is}{Ip} = \frac{Ep}{Es}$$

where Np = number of turns in primary winding
Ns = number of turns in secondary winding
Is = intensity of current in secondary winding
Ip = intensity of current in primary winding
Es = electromotive force (Volts) in secondary winding
Ep = electromotive force (Volts) in primary winding

EXAMPLE:
What will be the secondary voltage if the primary side is 14,400 V and the turns ratio is 60:1?

$$\frac{Np}{Ns} = \frac{Ep}{Es}$$

$$\frac{60}{1} = \frac{14{,}400\text{V}}{Es}$$

$$Es = \frac{14{,}400\text{ V}}{60}$$

$$= 240\text{ V}$$

PRACTICE EXERCISE: What will be the turns ratio if the primary voltage is reduced to 12,480 V and the secondary voltage remains the same?

2.12 TRANSFORMERS

2.12.1 Connecting the Delta Transformer

FROM THREE-PHASE SOURCE

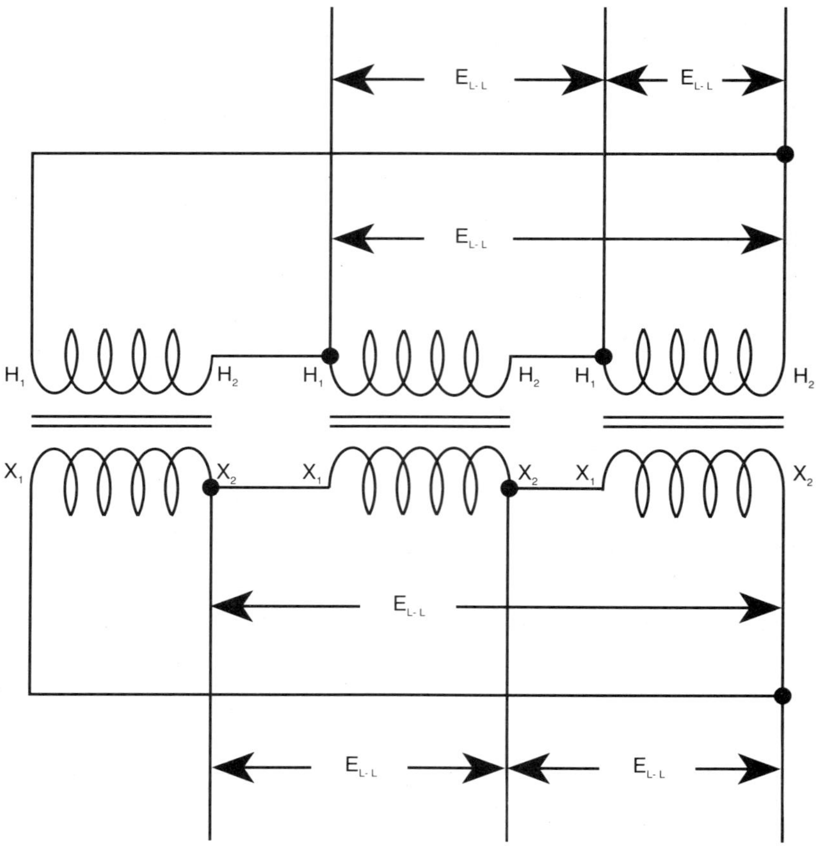

TO THREE-PHASE SECONDARY LOAD

$$E_{L-L} = Ecoil$$

$$Iline = Icoil \times 1.73 \text{ for balanced loads}$$

2.12.2 Power and Lighting Delta—Delta Connected Transformer Bank Unbalanced Loads

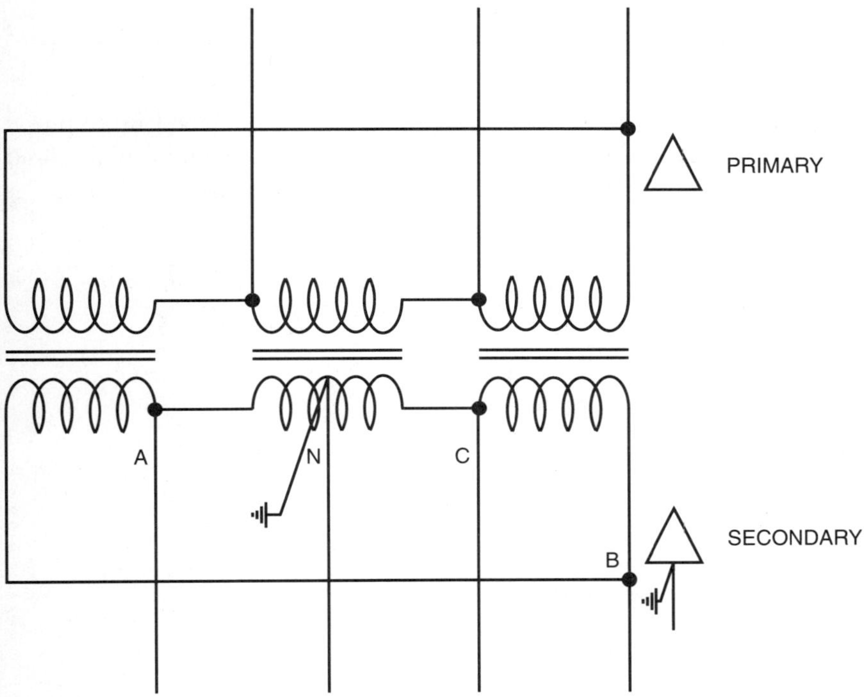

One secondary coil grounded mid-phase

$$\mathbf{I_T = Icoil_A + Icoil_B + Icoil_C}$$

$$\mathbf{I\text{ø}_A = \sqrt{I^2coil_B + I^2coil_C - 2Icoil_B \times Icoil_C \times \cos \angle A}}$$

$$\mathbf{I\text{ø}_B = \sqrt{I^2coil_A + I^2coil_C - 2Icoil_A \times Icoil_C \times \cos \angle B}}$$

$$\mathbf{I\text{ø}_C = \sqrt{I^2coil_A + I^2coil_B - 2Icoil_A \times Icoil_B \times \cos \angle C}}$$

$$\mathbf{\angle A = \angle B = \angle C = 180^\circ - 60^\circ \text{ for resistive circuits}}$$

$$\mathbf{E_{A-N} = E_{C-N} = \frac{E_{A-C}}{2}}$$

$$\mathbf{E_{B-N} = \frac{E_{A-C}}{2} \times \sqrt{3}}$$

IMPORTANT RULE: A center tap grounded transformer will carry approximately ⅔ of balanced line to neutral load across it, while the other two transformers will share equally the remaining balanced line to neutral load.

EXAMPLE:

A 120/240 V grounded delta system supplies a balanced three-phase load of 75 A across each transformer in addition to a single-phase load of 70 A across transformer coil B. What will be the phase currents?

$$
\begin{aligned}
I_A = I_C &= \sqrt{I^2coil_A + I^2coil_B - 2 \times Icoil \times Icoil \times \cos 120^\circ} \\
&= \sqrt{(75)^2 + (75+70)^2 - 2(75)(75+70)(-0.5)} \\
&= \sqrt{75^2 + 145^2 + (75)(145)} \\
&= \sqrt{26{,}650 + 10{,}875} \\
&= 193.7(194)\text{ A}
\end{aligned}
$$

$$
\begin{aligned}
I_B &= \sqrt{I^2coil_A + I^2coil_C - 2 \times Icoil_A \times Icoil_C \times \cos 120^\circ} \\
&= \sqrt{75^2 + 75^2 - 2(75)(75)(-0.5)} \\
&= \sqrt{75^2 + 75^2 + 75^2} \\
&= \sqrt{(3)(75)^2} \\
&= (\sqrt{3})(75) \\
&= 129.9(130)\text{ A}
\end{aligned}
$$

PRACTICE EXERCISE: What will be the phase currents from a transformer bank with 100 A across transformer coils A, B, and C with an additional load of 175 A across transformer coil B?

2.13 OPEN-DELTA (V CONNECTION)

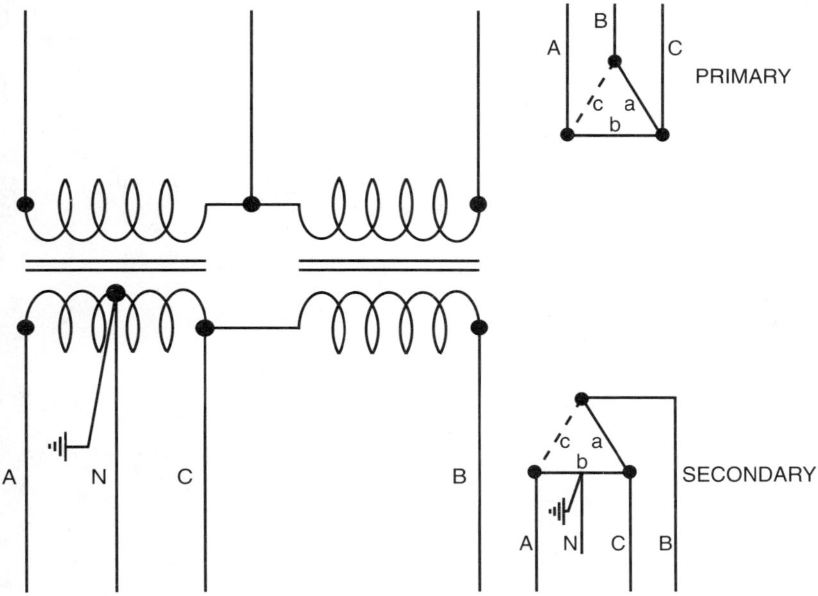

2.13.1 Voltages

$$E_{L\text{-}L} = Ecoil$$

$$E_{A\text{-}N} = E_{C\text{-}N} = \frac{E_{A\text{-}C}}{2}$$

$$E_{B\text{-}N} = \frac{E_{A\text{-}C}}{2}(1.73)$$

2.13.2 Currents

$$I_T = (Icoil_A + Icoil_B) \times (1.73)$$

$$I_A = Icoil_B$$

$$I_B = Icoil_A$$

$$I_C = \sqrt{I^2coil_A + I^2coil_B - 2Icoil_A \times Icoil_B \times \cos \angle C}$$

$$\angle C = (180° - 60°) \text{ for resistive loads}$$

2.14 POWER AND LIGHTING FROM A DELTA-WYE CONNECTED TRANSFORMER BANK

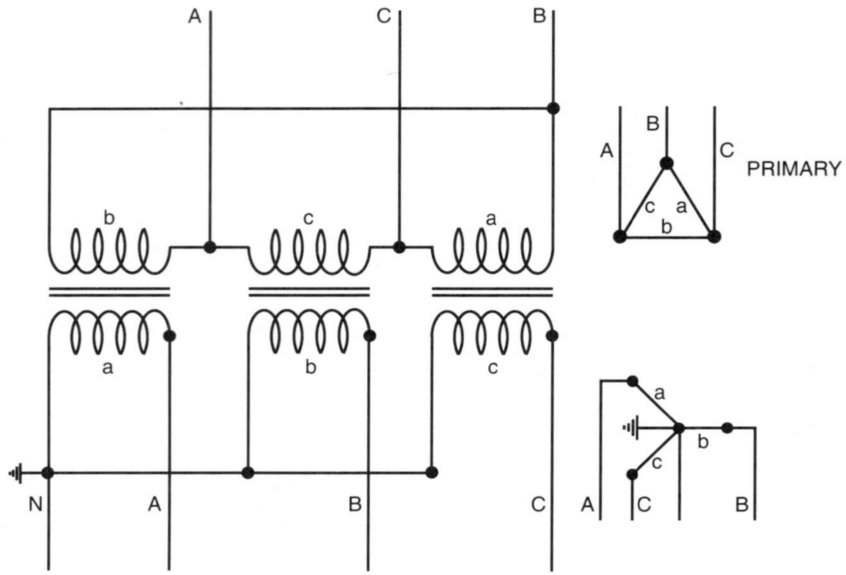

2.14.1 Voltages

Primary side: $\mathbf{E_{L-L} = Ecoil}$

Secondary side: $\mathbf{E_{L-L} = Ecoil(1.73)}$

$$\mathbf{E_{A-N} = E_{B-N} = E_{C-N} = Ecoil_A = Ecoil_B = Ecoil_C}$$

2.14.2 Currents

Primary side:

$$\mathbf{I_A = \sqrt{I^2coil_B + I^2coil_C - 2Icoil_B \times Icoil_C \times \cos\angle A}}$$

$$\mathbf{I_B = \sqrt{I^2coil_A + I^2coil_C - 2Icoil_A \times Icoil_C \times \cos\angle B}}$$

$$\mathbf{I_C = \sqrt{I^2coil_A + I^2coil_B - 2Icoil \times Icoil \times \cos\angle C}}$$

Secondary side: $\mathbf{I_A = Icoil_A}$ $\mathbf{I_B = Icoil_B}$ $\mathbf{I_C = Icoil_C}$

EXAMPLE:

What is the secondary side line-to-line voltage if the coil voltage is 277 V?

$$\begin{aligned} E_{L\text{-}L} &= \text{Ecoil}(1.73) \\ &= 277\ \text{V}(1.73) \\ &= 479.2(479)\ \text{V} \\ &= 480\text{V} \end{aligned}$$

PRACTICE EXERCISE: What is the current in the primary phase A if the secondary currents in phase B = 400 A and phase C = 800 A?

2.15 IMPEDANCE

$$\mathbf{Z = \sqrt{R^2 + X^2}}$$

EXAMPLE 1:

Determine Z when R = 1.98Ω, X_L = 1,885.0Ω, and X_C = 442.0Ω, and the resulting current is at 120 V pressure.

$$Z = \sqrt{(1.98)^2 + (1{,}885 - 442)^2 \Omega}$$
$$= \sqrt{(1.98)^2 + (1{,}443)^2 \Omega}$$
$$= \sqrt{1{,}433\ \Omega}$$
$$I = \frac{120\ \text{V}}{1{,}433\ \Omega}$$
$$= 0.083\ \text{A} = 83\ \text{mA}$$

EXAMPLE 2:

Determine current if L = C at 120 V.

$$Z = \sqrt{(1.98)^2 + (X_L - X_C)^2}$$
$$= \sqrt{(1.98)^2 + 0^2}$$
$$= 1.98\Omega$$
$$I = 120\ \text{V} \div 1.98\Omega = 60.6(61)\ \text{A}$$

PRACTICE EXERCISE: Determine Z when R = 10Ω, X_L = 65, X_C = 35 with current flow at 277 V pressure.

2.16 RESONANCE

When $X_L = X_C$, either deliberately or accidentally, reactance is cancelled and only resistance remains to oppose current flow.

EXAMPLE:

Consider the effect when $\omega L = \dfrac{1}{\omega C}$ in either series or parallel circuits, and when resistance is ignored where 5h is combined with 6µf:

$$2\pi f L = \frac{1}{2\pi f C}$$

$$f^2 = \frac{1}{4\pi^2 LC}$$

$$f = \frac{1}{2\pi\sqrt{LC}}$$

$$= \frac{1}{(6.28)[\sqrt{(5)(0.000006)}]}$$

$$= \frac{1}{(6.26)[\sqrt{(0.000030)}]}$$

$$= \frac{1}{0.034397}$$

$$= 29 \text{ Hz}$$

PRACTICE EXERCISE: Determine the resonating frequency when $X_L = 35\Omega$ and $X_C = 65\Omega$.

2.17 SERIES CIRCUITS HAVING RESISTANCE, INDUCTANCE, AND IMPEDANCE

Resistance and Inductance

$$\mathbf{Z = \sqrt{R^2 + X_L^2}}$$
$$\mathbf{E_T = \sqrt{E_R^2 + E_L^2}}$$

Resistance and Capacitance

$$\mathbf{Z = \sqrt{R^2 + X_C^2}}$$
$$\mathbf{E_T = \sqrt{E_R^2 + E_C^2}}$$

Resistance, Inductance, and Capacitance

$$\mathbf{Z = \sqrt{R^2 + (X_L - X_C)^2}}$$
$$\mathbf{E_T = \sqrt{E_R^2 + E^2_{X_L - X_C}}}$$

EXAMPLE:

Determine the impedance of a circuit that has 25 ohm resistance, 40 ohm inductive reactance, and 10 ohm capacitive reactance.

$$\begin{aligned} Z &= \sqrt{R^2 + (X_L - X_C)^2} \\ &= \sqrt{25^2 + (40\Omega - 10\Omega)^2} \\ &= \sqrt{625 + 900\Omega^2} \\ &= \sqrt{1{,}525\Omega^2} \\ &= 39\Omega \end{aligned}$$

PRACTICE EXERCISE: What is the impedance of this circuit if the resistance drops to zero?

2.18 PARALLEL CIRCUITS HAVING RESISTANCE, INDUCTANCE, AND IMPEDANCE

Resistance and Inductance

$$\mathbf{I_T = \sqrt{I_R^2 + I_L^2}}$$

Resistance and Capacitance

$$\mathbf{I_T = \sqrt{I_R^2 + I_C^2}}$$

Resistance, Inductance, Capacitance

$$\mathbf{I_T = \sqrt{I_R^2 + (I_L - I_C)^2}}$$

Impedance

$$\mathbf{Z = E/I_T}$$

where $I_R = \dfrac{E}{R}$

$$I_L = \frac{E}{X_L}$$

$$I_C = \frac{E}{X_C}$$

EXAMPLE:

What is the apparent impedance of a circuit when R = 10Ω, L = 83 mh, C = 100μf at 120 V and 60 Hz?

$$\begin{aligned} X_L &= 2\pi f L = \omega L \\ &= (377)(0.083\text{h}) \\ &= 31.2 \text{ ohms} \\ X_C &= \frac{1}{2\pi f C} \\ &= \frac{1}{\omega C} \\ &= \frac{1}{377 \times 0.0001\text{f}} \\ &= \frac{1}{0.0377} \end{aligned}$$

$$= 26.5 \text{ ohms}$$

$$I_R = \frac{120 \text{ V}}{10\Omega}$$

$$= 12 \text{ A}$$

$$I_L = \frac{120 \text{ V}}{31.2\Omega}$$

$$= 3.85 \text{ A}$$

$$I_C = \frac{120 \text{ V}}{26.5\Omega}$$

$$= 4.5 \text{ A}$$

$$I_T = \sqrt{I_R^2 + (I_L - I_C)^2}$$

$$= \sqrt{12^2 + (3.85 - 4.5)^2}$$

$$= \sqrt{12^2 + 0.4225^2}$$

$$= \sqrt{144.4}$$

$$= 12 \text{ A}$$

PRACTICE EXERCISE: What is the apparent impedance of a circuit when R = 5Ω, L = 100 mh, C = 50zμf at 120 V and 60 Hz?

2.19 POWER FACTOR

Apparent power = V × I or VA

For single-phase: VA × 1

For three-phase: VA × 1.73

EXAMPLE:

What is the power factor (pf) of a circuit with output load of 9,600 W and input power of 100 A at 120 V?

$$\begin{aligned}\text{pf} &= \frac{\text{true power}}{\text{apparent power}}\\ &= \frac{9{,}600\ \text{W}}{100\ \text{A} \times 120\ \text{V}}\\ &= \frac{9{,}600\ \text{W}}{12{,}000\ \text{VA}}\\ &= 0.8\\ &= 80\%\end{aligned}$$

PRACTICE EXERCISE: What is the apparent power if the pf = 65% and the load wattage is 10 kW?

2.20 POWER FACTOR CORRECTION

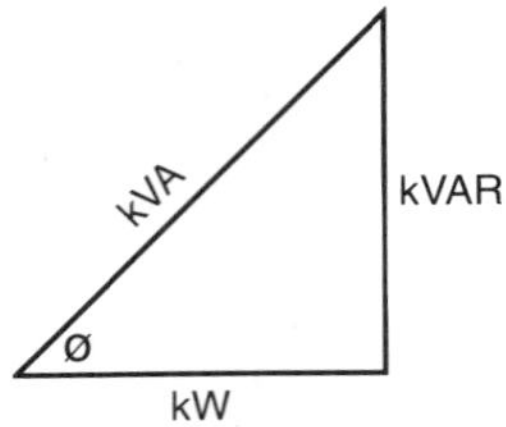

$$\mathbf{kVA} = \sqrt{\mathbf{kW}^2 + \mathbf{kVAR}^2}$$
$$\mathbf{kW} = \mathbf{kVA}\ \mathbf{cos}\ \mathbf{Ø}$$
$$\mathbf{kVARS} = \mathbf{kVA}\ \mathbf{sin}\ \mathbf{Ø}$$

EXAMPLE:

A 25 hp three-phase motor operates at 460 V and 75% power factor (pf). Correct to no more than 10% current reduction from the 34 A shown in *Table 430-150.*

$$\text{Original kVA} = \frac{(34\text{ A})(460\text{ V})(1.73)}{\frac{1{,}000\text{ VA}}{\text{kVA}}}$$
$$= (34\text{ A})(460\text{ V})(1.73) \times \frac{\text{kVA}}{1{,}000\text{ VA}}$$
$$= 27.05(27)$$
$$\text{Original kW} = (27\text{ kVA})(0.75)$$
$$= 20.25(20)$$
$$\text{Original kVAR} = (27\text{ kVA})(\text{sinArccos } 0.75)$$
$$= 17.85(18)$$

$$\text{Minimum new current} = (90\%)(34\text{ A})$$
$$= 30.6(31)\text{ A}$$

$$\text{New kVA} = \frac{(31\text{ A})(460\text{ V})(1.73)}{1{,}000\text{ VA/kVA}}$$
$$= 24.6\,(25)$$

$$\text{New pf} = \text{kW/kVA} = 20/25 = 0.8 = \cos \text{Ø}$$

$$\text{New kVARS} = \text{kVA(sinArccos 0.8)} = (25\ \text{kVA})(0.6) = 15$$

$$\text{Minimum kVARS needed} = (18 - 15) = 3$$

$$\text{New FLA} = \frac{20\ \text{kW}}{(0.8)(.460\ \text{kV})(1.73)} = 31.4\ (31)\ \text{A}$$

$$\text{Actual percentage of current reduction} = \frac{(34 - 31)\ \text{A}}{34\ \text{A}} = 0.088\ (0.09) = 9\%$$

PRACTICE EXERCISE: Determine the resultant current of a 15 hp 230 V three-phase motor with 80% pf corrected for no more than 10%.

2.20.1 Power Factor Correction Short-Cut

EXAMPLE 1:

When kW, kVAR, and original power factor (pf) are known, determine the ckVAR necessary to improve the pf.

$$\text{ckVAR} = \text{kW}(\tan 1 - \tan 2)$$

where ckVAR = capacitance in kVAR

tan 1 = tangent of original angle

tan 2 = tangent of final angle

EXAMPLE 2:

With 20 kW and 18 kVAR at 75% pf, how much ckVAR is needed to improve pf to 95%?

$$\begin{aligned} \text{ckVAR} &= 20\,(\tan \text{Arcos}\ 0.75 - \tan \text{Arcos}\ 0.95) \\ &= 20\,(0.8819 - 0.3286) \\ &= 11.0 \end{aligned}$$

PRACTICE EXERCISE: How much ckVAR is needed to correct to 90% pf when the original load is 30 kVA at 70% pf?

2.21 CONDUCTOR RESISTANCE CONVERSION

$$\mathbf{R_F = R_o\,[1 + \alpha\,(T_2 - 75)]}$$

where R_F = resultant conductor resistance at new temperature

R_o = conductor resistance at 75° C

$\alpha = 0.00323$ for copper at 75° C

T_2 = new temperature

EXAMPLE:

What is the resistance of #12 AWG copper-stranded conductor per 1,000 feet and its ohms-circle mil per foot value at 50° C in an industrial facility?

$$\begin{aligned} R_F &= R_o\,[1 + \alpha(T_2 - 75)] \\ &= (1.98\Omega \div 1{,}000 \text{ ft}) \times [1 + 0.00323(50° - 75°)] \\ &= (1.98\Omega \div 1{,}000 \text{ ft}) \times [1 + (-0.08075)] \\ &= \frac{1.82 \text{ ohms}}{1{,}000 \text{ feet}} \end{aligned}$$

$$\begin{aligned} \frac{\text{ohms–circle mil}}{\text{foot}} &= \frac{(1.82 \text{ ohms})(6{,}530 \text{ circle mil})}{1{,}000 \text{ feet}} \\ &= \frac{11.88 \text{ ohms–circle mil}}{\text{foot}} \end{aligned}$$

PRACTICE EXERCISE: What is its ohms-circle mil per foot of the Example at 20° C?

CHAPTER 3

Branch Circuits

3.1 NON-CONTINUOUS NON-HARMONIC LOAD

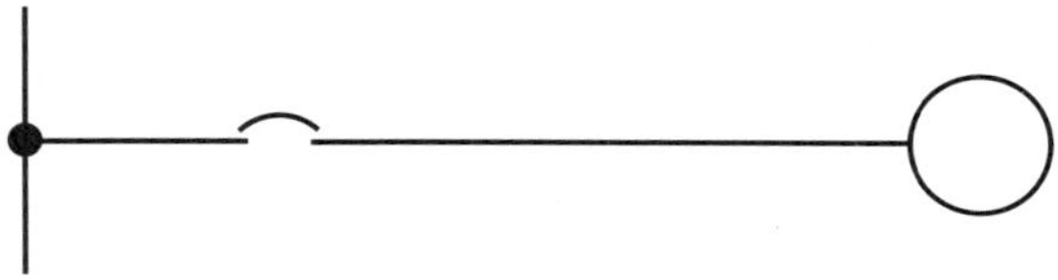

In this chapter and Chapter 4, references to the *1993 National Electrical Code®* are designated by the ► symbol. It is recommended that you have the *NEC®* book readily available while following the examples given.

3.1.1 Single Loads

► Branch circuits supplying a single load can be any size necessary (*Section 210-3*).

EXAMPLE:

We have a 60-amp non-continuous load with terminals and enclosure marked suitable for use with 60/75° C conductor ampacity, copper only, in an ambient not greater than 30° C. An insulated equipment grounding conductor will be used with all circuits and insulations will be rated at 90° C.

► Minimum wire size: Use *Sections 110-14*, *210-19*, and *310-15(a),* select #6 with 90° C insulation.

► Minimum protection (fuse or circuit breaker): Use *Section 210-20*, select 60 amperes.

► Minimum equipment grounding conductor: Use *Section* and *Table 250-95*, select #10 AWG.

► Minimum neutral circuit conductor: Use *Section* and *Table 210-19,* select #6 AWG.

- ▶ Minimum raceway: Use *Section 300-17.*
- ▶ Single-phase two-wire with ground: Use Chapter 9 *Tables 4* and *5*, select #6 AWG.
  [(2)(0.0519) + (1)(0.0184)] = 0.1218; use 3/4 inch
- ▶ Single-phase three-wire with ground: Use Chapter 9 *Tables 4* and *5*, select #6 AWG.
  [(3)(0.0519) + (1)(0.0184)] = 0.1741; use 3/4 inch
- ▶ Three-phase four-wire with ground: Use Chapter 9 *Tables 4* and *5*, select #6 AWG.
  [(4)(0.0519) + (1)(0.0184)] = 0.2260; use 1 inch
- ▶ DC resistance and voltage drop for a 100-foot long circuit: Use Chapter 9 *Table 8.*

At 75° C:

$$\frac{0.491\Omega}{1{,}000 \text{ ft}} \times 100 \text{ ft} = 0.0491\Omega$$

$$2(0.0491\Omega)(60\text{A}) = 5.892(6) \text{ V for single-phase}$$

$$(6 \text{ V})(0.866) = 5.196(5) \text{ V for three-phase}$$

At 20° C:

$$R_F = R_o\,[1 + (T_2 - T_1)] \quad \text{where } \alpha = 0.00323, \; T_1 = 75, \; T_2 = 20$$

$$= \frac{(0.491\Omega)}{1{,}000 \text{ ft}} \times [1 + (0.00323)(20 - 75)]$$

$$= \frac{(0.491\Omega)}{1{,}000 \text{ ft}} \times (1 - 0.17765)$$

$$= \frac{0.4038\Omega}{1{,}000 \text{ ft}}$$

$$\frac{(0.4038\Omega)}{1{,}000 \text{ ft}} \times (100 \text{ ft}) = 0.04038\ \Omega$$

$$(2)(0.04038\Omega)(60\text{A}) = 4.846(5) \text{ V for single-phase}$$

$$(5 \text{ v})(0.866) = 4.33(4) \text{ V for three-phase}$$

PRACTICE EXERCISE: Consider a 100-amp non-continuous load with terminals and enclosure marked suitable for use with 60/75° C conductor ampacity, copper only, in an ambient temperature not greater than 30° C. An insulated equipment grounding conductor will be used with all circuits, and all insulations will be rated 90° C. Determine the minimum wire size, protection, equipment grounding conductor, raceway, and DC resistance and voltage drop for a three-phase four-wire circuit 175 feet in length located within a 40° C environment.

3.2 CONTINUOUS NON-HARMONIC LOAD

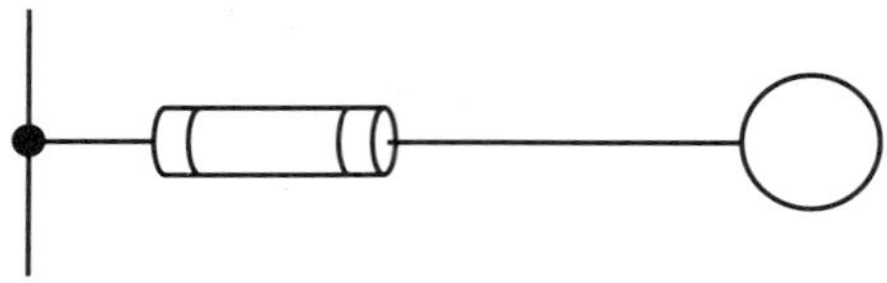

3.2.1 Single Loads

► Branch circuits supplying a single load can be any size necessary (*Section 210-3*).

EXAMPLE:

Consider a 60-amp continuous load with terminals and enclosure marked suitable for use with 60/75° C conductor ampacity, copper only, in an ambient not greater than 30° C. An insulated equipment grounding conductor will be used with all circuits and insulations will be 90° C.

► Minimum wire size: Use *Sections 110-14, 210-19*, and *310-15(a)*, select #6 with 90° C insulation.

► Initial wire size is #6 AWG (*Section 210-19*).
Minimum protection based on continuous current through the overcurrent devices (60 A)(125%) = 75 A; round up to 80 A. Since the #6 AWG conductor 90° C insulation rating is 75 A, it is then protected at its ampacity.

► (*Sections 210-22(c)* and *240-3* Ex. 4)

► Minimum equipment grounding conductor: Use *Section* and *Table 250-95*, select #10 AWG.

► Minimum neutral circuit conductor: Use *Section* and *Table 210-19*, select #6 AWG.

► Minimum raceway: Use *Section 300-17*.

► Single-phase two-wire with ground: Use Chapter 9 *Tables 4* and *5*, select #6 AWG.
[(2)(0.0519) + (1)(0.0184)] = 0.1218; use 3/4 inch

► Single-phase three-wire with ground: Use Chapter 9 *Tables 4* and *5*, select #6 AWG.
[(3)(0.0519) + (1)(0.0184)] = 0.1741; use 3/4 inch.

- Three-phase three-wire with ground: Use Chapter 9 *Tables 4* and *5*, select #6 AWG.
  [(3)(0.0519) + (1)(0.0184)] = 0.1741; use 3/4 inch.
- Three-phase four-wire with ground: Use Chapter 9 *Tables 4* and *5*, select #6 AWG.
  [(4)(0.0519) + (1)(0.0184)] = 0.2260; use 1 inch.
- DC resistance and voltage drop for a 100-foot long circuit: Use Chapter 9 *Table 8*.

At 75° C:

$$\frac{0.491\Omega}{1{,}000\text{ ft}} \times 100\text{ ft} = 0.0491\Omega$$

$$2(0.0491\Omega)(60\text{A}) = 5.892(6)\text{ V for single-phase}$$

$$(6\text{ V})(0.866) = 5.196(5)\text{ V for three-phase}$$

At 20° C:

$$R_F = R_o\,[1 + (T_2 - T_1)] \quad \text{where } \alpha = 0.00323$$

$$T_1 = 75$$

$$T_2 = 20$$

$$= \frac{(0.491\Omega)}{1{,}000\text{ ft}} \times [1 + (0.00323)(20 - 75)]$$

$$= \frac{(0.491\Omega)}{1{,}000\text{ ft}} \times (1 - 0.17765)$$

$$= \frac{0.4038\ \Omega}{1{,}000\text{ ft}}$$

$$\frac{(0.4038\Omega)}{1{,}000\text{ ft}} \times (100\text{ ft}) = 0.04038\Omega$$

$$(2)(0.04038\Omega)(60\text{A}) = 4.846(5)\text{ V for single-phase}$$

$$(5\text{ v})(0.866) = 4.33(4)\text{ V for three-phase}$$

PRACTICE EXERCISE: Consider a 100-amp continuous load with terminals and enclosure marked suitable for use with 60/75° C conductor ampacity, copper only, in an ambient temperature not greater than 30° C. An insulated equipment grounding conductor will be used with all circuits and all insulations will be rated 90° C. Determine the minimum wire size, protection, equipment grounding conductor, raceway, and DC resistance and voltage drop for a three-phase four-wire circuit 175 feet in length located within a 40° C environment.

3.3 HARMONIC BRANCH CIRCUIT LOADS

3.3.1 Single Loads

► Branch circuits supplying a single load can be any size necessary (*Section 210-3*).

EXAMPLE 1:

A non-linear lighting load is supplied from a wye-connected source having neutral currents sharing a conductor common to 2 or 3 phases. The electric discharge fixtures contain 2 ballasts rated at 0.72 A each at 120 V and supplied by 20 A circuit breaker. Terminals and enclosures are marked suitable for use with 60/75° C conductor ampacity, copper only, in an ambient not greater than 30° C. An insulated equipment grounding conductor will be used with all circuits, and insulations will be 90° C (*Section 110-14*).

Current demand of each fixture:

$$(0.72\text{ A})(2) = 1.44\text{ A}$$

$$(1.44\text{ A})(120\text{ V}) = 172.8(173)\text{ VA}$$

Maximum VA of OC device:

$$(20\text{ A})(120\text{ V}) = 2{,}400\text{ VA}$$

► Maximum continuous load across OC device: Use *Section 210-22(b).*

$$\frac{(2{,}400\text{ VA})}{(1.25)} = 1{,}920\text{ VA}$$

► Maximum number of fixtures permitted: Use *Section 210-22(c).*

$$\frac{1{,}920\text{ VA}}{\frac{173\text{ VA}}{\text{fixture}}} = 1{,}920\text{ VA} \times \frac{\text{fixture}}{173\text{ VA}} = 11.09(11)\text{ fixtures}$$

► Minimum 90° C rated conductor selected is #12 AWG: Use *Section 310-15(a).*

► Use *Tables 310-16* through *19* and Notes 8(a) and 10(c). The maximum current rating with 4 heat-producing conductors in a raceway or cable is:

$$(30\text{ A})(80\%) = 24\text{ A}$$

Recommended phase conductor sizing technique:

(0.72 A)(2)(1.25)(11) = 19.8(20) A, which is less than its derated ampacity. (Recommended neutral conductor size for a two-wire branch circuit is that it be the same size as the phase conductor.)

Recommended neutral conductor sizing technique when it is common to the three phases of a wye system:

(0.72 A)(2)(1.73)(11) = 27.4(27) A, which is greater than its derated ampacity. Increase the common conductor to #10 AWG and refigure:

(40 A)(80%) = 32 A, which is greater than the circuit load current and is acceptable.

If only two phases are energized, and ignoring individual phase inductive lag angle:

$$
\begin{aligned}
N &= \sqrt{a^2 + c^2 - (2)(a)(c)\cos 60^\circ}\ \text{A} \\
&= \sqrt{27^2 + 27^2 - 2(27)(27)(0.5)}\ \text{A} \\
&= \sqrt{27^2 + 27^2 - (27)^2}\ \text{A} \\
&= \sqrt{27^2}\ \text{A} \\
&= 27\ \text{A}
\end{aligned}
$$

Recommended sizing of the neutral common conductor for the preceding arrangement with two phases energized: (27 A)(1.25) = 33.75(34) A, which is greater than the derated ampacity of #10 AWG and should be further increased to #8 AWG:

> (55 A)(80%) = 44 A, which is greater than the neutral load current for this arrangement, and is acceptable.

PRACTICE EXERCISE: A three-phase four-wire circuit from a wye source will supply receptacle outlets supplying office copiers, computers, and computer terminals throughout the area. An equipment grounding conductor is included and only five receptacles will be on each circuit. Determine the circuit rating and conductor sizes when all terminals are rated for 60° C and insulated with 90° C material, with an equipment grounding conductor, and minimum raceway needed.

3.4 ELECTRIC RANGES

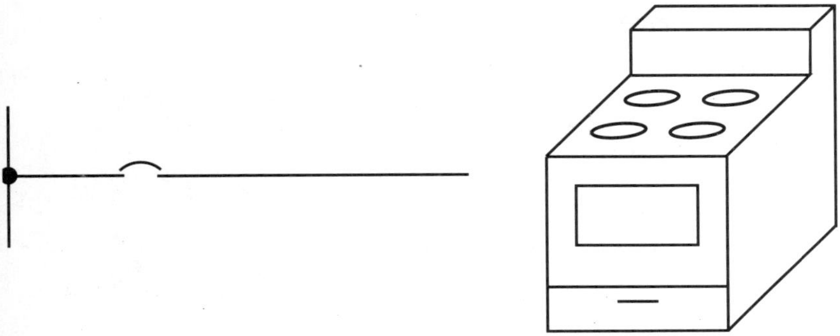

EXAMPLE:

Single range rated 12.5 kW

► Use *Section 220-3(c)* Exception 2:

(12.5 - 12.0) kW = 0.5 kW, round up to 1 kW

$1 \text{ kW} \times \frac{5\%}{\text{kW}} = 5\%$; increase the demand factor by this amount.

$$[(5\%)\ (8\text{kW}) + 8\text{kW}]$$
$$= 8.4 \text{ kW}$$
$$= 8.4 \text{ kVA}$$

$$\frac{8.4 \text{ kVA}}{0.240 \text{ kV}} = 35 \text{ A}$$

► Use *Section 336-26*, select #8 AWG.

► Minimum overcurrent device = 40 A. Use *Section 210-19(b)*. Minimum neutral circuit conductor: (40 A)(70%) = 28 A.

► Use #10 AWG *Section 210-19(b)* Exception 2, select #10 AWG. When the neutral is also used as an equipment grounding conductor, *Section 250-60* requires at least a #10 AWG.

PRACTICE EXERCISE: A single range with a nameplate rating of 15 kW is supplied by a cable marked with NMC-B and the source is 120/240 V. Determine the circuit rating and conductor size for the ungrounded and grounded wires.

3.5 COOKTOP

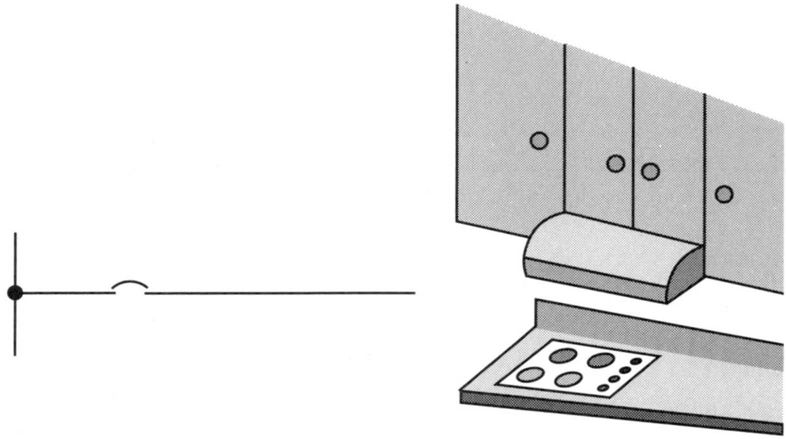

EXAMPLE 1:

Cooktop rated 6 kW

► Use *Section 210-19(b)* and *Table 220-19* Note 4:

$$6\text{ kW} = \frac{6\text{ kVA}}{0.240\text{ kV}} = 25\text{ A, use \#10 AWG.}$$

Overcurrent device: Use *Section 422-28(e).*
(25 A)(150%) = 37.5(38) A; initial selection could be 40 A and #8 AWG, while the minimum could be as low as 25 A, permitting the use of the #10 AWG.

Minimum neutral conductor: Use #10 AWG.

3.6 WALL OVEN

EXAMPLE:

Wall oven rated 4.5 kW

► Use *Section 210-19(b)* and *Table 220-19* Note 4:

$$4.5\text{ kW} = \frac{4.5\text{ kVA}}{0.24\text{ kV}} = 18.75(19)\text{ A; use \#12 AWG}$$

► Overcurrent device: Use *Section 422-28(e).*
Maximum: (19 A)(150%) = 28.8(29) A; use 30 A and #10 AWG.
Minimum: as low as 20 A, permitting the use of the #12 AWG.

► Minimum neutral conductor size would be the same size as the ungrounded conductors. Use *Section 210-19(b)* Exception 2.

3.7 COOKTOP AND WALL OVENS

EXAMPLE:

6 kW cooktop and 2 4.5 kW wall ovens in the same room

▶ Use *Section 210-19(b)* and *Table 220-19* Note 4:

$$(6\text{ kW} + 4.5\text{ kW} + 4.5\text{ kW}) = (15\text{ kW} - 12\text{ kW}) = 3\text{ kW}$$

$$(3\text{ kW})\,\frac{(5\%)}{\text{kW}} = [(15\%)(8\text{ kW}) + 8\text{ kW}]$$

$$= 9.2\text{ kW}$$

$$= 9.2\text{ kVA}$$

$$\frac{9.2\text{ kVA}}{0.240\text{ kV}} = 38.3(38)\text{ A;}$$

Overcurrent device:

▶ Use *Section 210-19(b)*.
Minimum: 40 A
Neutral conductor:

▶ Use *Section 210-19(b)* Exception 2.
Minimum: (40A)(70%) = 28A; use #10 AWG.

PRACTICE EXERCISE: A 4 kW cooktop and a 5 kW wall oven are supplied by a cable with NMC-B markings, and both are located in the same room. Determine the circuit rating and conductor size.

3.8 DWELLING LIGHTING CIRCUITS

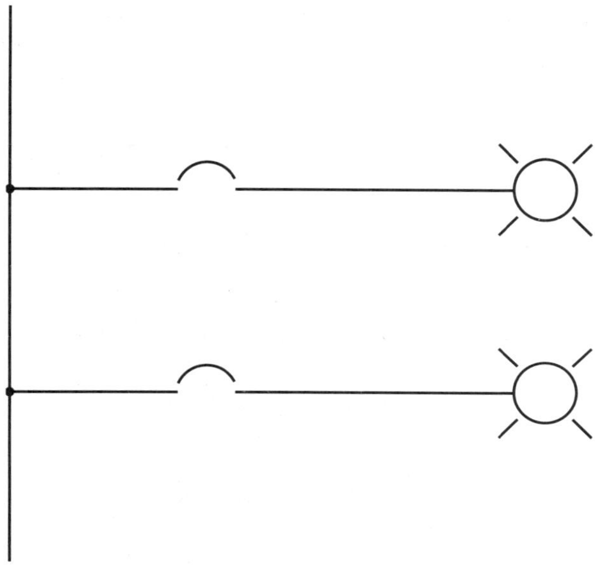

EXAMPLE:

A 1,200-ft^2 house, exclusive of the open areas or garage

► Minimum lighting circuits: Use *Section 220-3(b).*

$$(1{,}200\ ft^2)\frac{(3\ VA)}{ft^2} = 3{,}600\ VA$$

$$\frac{3{,}600\ VA}{120\ V} = 30\ A$$

$$\frac{30\ A}{\frac{15\ A}{circuit}} = \frac{30\ A}{15\ A} \times circuits$$

$$= 2\ circuits\ with\ \#14\ AWG,\ or$$

$$\frac{30\ A}{\frac{20\ A}{circuit}} = \frac{30\ A}{20\ A} \times circuit$$

$$= 1.5(2)\ circuits\ with\ \#12\ AWG$$

PRACTICE EXERCISE: An 1,800-square-foot dwelling, exclusive of open areas and garage, requires how many 15-ampere circuits? How many 20-ampere circuits?

3.9 ELECTRIC HEATING

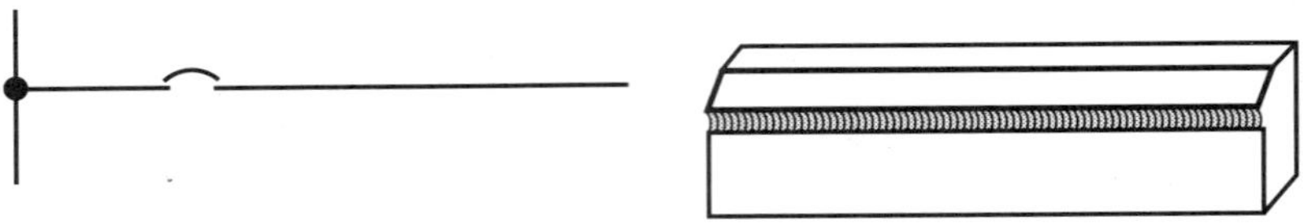

EXAMPLE:

Non-central heating is separately thermostatically controlled, with one 500-watt, ten 1,000-watt and one 1,500-watt units installed at various locations in a dwelling and supplied with 240 V single-phase.

▶ Minimum number circuits for 15 A and #14 AWG; 20 A and #12 AWG; or 30 A and #10 AWG (*Section 424-3*).

15 A circuit = (240 V)(15 A)(80%) = 2,880 VA
= 2,880 W

500 W plus 1,500 W units for one circuit and 5 circuits of 2 1,000 W units each, for a total of 6 circuits.

20 A circuit = (240 V)(20 A)(80%) = 3,840 VA
= 3,840 W

500 W plus 3 1,000 W units for one circuit, 1,500 W plus 2 1000 W units for one circuit, 3 1,000 W units for one circuit, and 2 1,000 W units for one circuit, for a total of 4 circuits.

30 A circuit = (240 V)(30 A)(80%) = 5,760 VA
= 5,760 W

500 W plus 5 1,000 W units for one circuit, 1,500 plus 4 1,000 W units for one circuit, 1 1,000 W unit for one circuit, for a total of 3 circuits.

▶ Other combinations can be used as long as the load does not exceed 80% of the branch circuit rating, or the maximum permitted by the listing or labeling of the heaters (*Section 110-3*).

PRACTICE EXERCISE: How many individual baseboard wall heater units rated 1,000 watts at 240 V can be placed on a 30 A branch circuit? They are separately controlled. What is the maximum if the source is 208 V?

3.10 BANK/OFFICE BUILDING BRANCH CIRCUITS

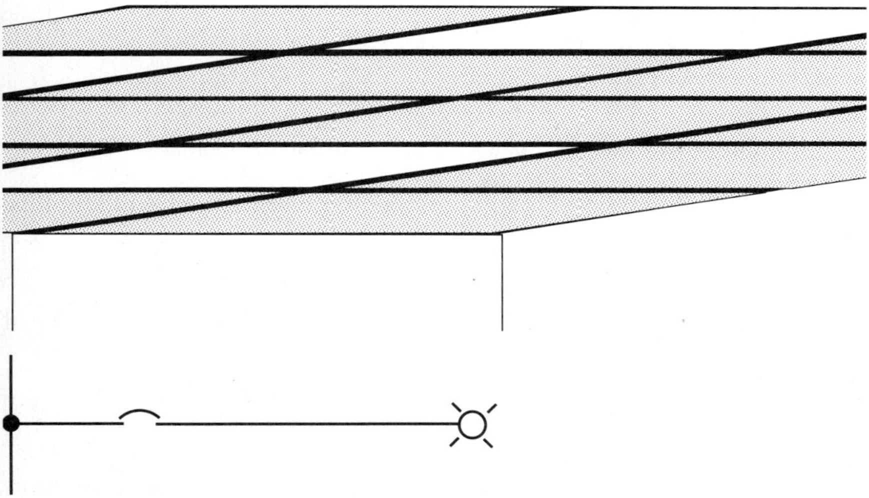

EXAMPLE:

Five-story 40 ft × 60 ft building with 10 ft × 10 ft stairwell and 10 ft × 10 ft elevator areas provided. Assuming continuous loads and harmonic currents present operating at 120 V and from a three-phase wye source, determine the number of branch circuits necessary when approximately 50% of each 20A circuit is reserved for future growth.

Total square foot area for general lighting:

$$(5)[(40\text{ ft})(60\text{ ft}) - (10\text{ ft})(10\text{ ft}) - (10\text{ ft})(10\text{ ft})] = 11{,}000\text{ ft}^2$$

$$(11{,}000\text{ ft}^2)\frac{(3.5\text{ VA})}{\text{ft}^2} = 38{,}500\text{ VA}$$

$$\frac{38{,}500\text{ VA}}{\frac{120\text{ V}}{\text{phase}}} = \frac{38{,}500\text{ VA}}{120\text{ V}} \times \text{phases}$$

$$= \frac{38{,}500\text{ VA} \times \text{phases}}{120\text{ V} \times 3\text{ phases}}$$

$$= \frac{38{,}500\text{ A}}{(120)(3)}$$

$$= 107 \text{ A–phase}$$

$$\frac{107 \text{ A–phase}}{\frac{10 \text{ A}}{\text{pole}}} = \frac{107 \text{ A–phase}}{10 \text{ A}} \times \text{pole}$$

$$= 10.7(11) \text{ pole each phase}$$

Minimum recommended phase conductor load allowance:

$\frac{107 \text{ A}}{11} = \frac{9.72(10) \text{ A}}{\text{phase}} = \frac{(10 \text{ A})(1.25)}{\text{phase}} = \frac{12.5(13) \text{ A}}{\text{phase}}$; which is less than the derated ampacity of 90° C #12 AWG.

Minimum recommended neutral circuit conductor size:

(10 A)(1.73) = 17.3(17) A; which is less than the derated ampacity of 90° C #12 AWG.

► Use *Tables 310-16* through *19* and Notes 8(a) and 10(c).

PRACTICE EXERCISE: How many branch circuits are necessary for a 60 ft × 60 ft office building where the actual connected lighting load is 10,000 VA in addition to 60 duplex receptacle outlets for harmonic loads such as copiers, computers, and terminals? The supply is from a 208Y/120 source.

3.11 GENERAL PURPOSE RECEPTACLES, LOAD ALLOWANCES

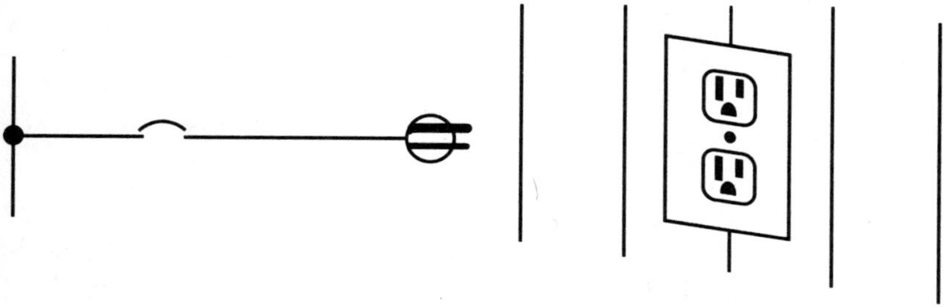

► General purpose receptacle circuits require 180 VA minimum capacity. Use *Section 220-3(c)(6)*:

$$(15\ \text{A})(120\ \text{V}) = \frac{1{,}800\ \text{VA}}{\dfrac{180\ \text{VA}}{\text{Receptacle}}}$$

$$= \frac{1{,}800\ \text{VA}}{180\ \text{VA}} \times \text{Receptacles}$$

$$= 10\ \text{Receptacles}$$

$$(20\ \text{A})(120\ \text{V}) = \frac{2{,}400\ \text{VA}}{\dfrac{180\ \text{VA}}{\text{Receptacles}}}$$

$$= \frac{2{,}400\ \text{VA}}{180\ \text{VA}} \times \text{Receptacles}$$

$$= 13.3(13)\ \text{Receptacles}$$

3.12 MOTOR BRANCH-CIRCUIT

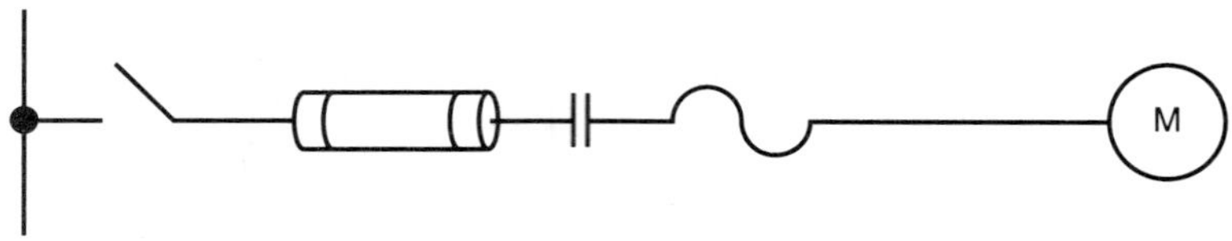

EXAMPLE:

A 460 V 3-phase 60 hp motor has a Code Letter F, 40° C rise and 1.0 SF on its nameplate. The combination enclosure used is suitable for 75° C conductor ampacity, but the controller (starter) is suitable only for copper with conductors sized from the 60° C column of *Table 310-16.*

3.12.1 Overcurrent Protection Calculations

- Motor full load current (FLA): Use *Section 430-6* and *Table 310-16.* 77 A.

- Minimum conductor size: Use *Sections 110-3(a)* and *(b), 430-22(a),* and *Table 310-16.*
  (77 A)(125%) = 96.25(96) A; use #1 AWG.

- Motor locked rotor current: Use *Section* and *Table 430-7(b).*

$$\frac{(5.59 \text{ kVA})(60 \text{ hp})}{(\text{hp})(0.460\text{kV})(1.73)} = 421.4(421) \text{ A}$$

- Overload protection: Use *Sections 430-32(a)(1)* and *430-34.*
  Minimum: (77 A)(125%) = 96.25 A
  Maximum: (77 A)(140%) = 107.8 A

- Overcurrent Protection: Use *Section 430-52* and *Table 430-152.*
  Non-time-delay fuse:
  (300%)(77 A) = 231; use 225 A.

  Time-delay fuse:
  (175%)(77 A) = 134.75; use 125 A.

  Time-delay fuse ten second rating:
  (5)(125 A) = 625 A, which exceeds the calculated LRA of this motor.

Inverse time circuit breaker:
(250%)(77 A) = 192.5 A; use 175 A.

Instantaneous circuit breaker:
(700%)(77 A) = 539 A; use 500 A setting.

► Instantaneous circuit breaker minimum handle rating: Use *Section 430-110(a)*.
(77 A)(115%) = 88.55; use 90 A.

Assume, however, that this motor has an LRA code letter K instead.

$$\frac{(8.99\ \text{kVA})(60\ \text{HP})}{(\text{HP})(0.460\ \text{V})(1.73)} = 677.8(678)\ \text{A}$$

Since this LRA exceeds time-delay rating of the first TD fuse selected, it will be necessary to round up: time delay would then be (5)(150 A) = 750 A. The instantaneous trip circuit breaker setting will need to be adjusted for this higher value as well.

In addition, if the motor has a lengthy start-up time before coming to full speed, rounding up may be needed with lower LRA values as well, and a time-delay bypass of the running overload protection may be necessary.

► Use as permitted per *Section 430-35*.

► Equipment grounding conductor: Use *Section 250-95*, Exceptions and Table.
Minimum with the initial fuses selected.
Non-time-delay: Use #4 AWG.
Time-delay fuses: Use #6 AWG.
Inverse time circuit breaker: Use #6 AWG.
Instantaneous trip circuit breakers: #8 or #6 AWG depending upon the overload device rating.

3.12.2 Raceway Sizing

► Minimum raceway size (*Section 300-17*):
Using a #4 AWG equipment grounding wire:
TBL 5 & 4 [(3)(0.1590) + (1)(0.0845)] = 0.5615; use 1-1/4 inch.

Using a #6 AWG equipment grounding wire:
TBL 5 & 4 [(3)(0.1590) + (1)(0.0519)] = 0.5289; use 1-1/4 inch.

Using a #8 AWG equipment grounding wire:
TBL 5 & 4 [(3)(0.1590) + (1)(0.0373)] = 0.5143; use 1-1/4 inch.

3.12.3 Voltage Drop

► Use Chapter 9, Table 8

Assume an industrial environment, 75 feet one way with 40° C ambient temperature. The resultant ampacity based on the elevated temperature for #1 AWG @ 90° C = (150 A)(0.91) = 136.5(137) A; greater than the 96.25 A minimum required and still acceptable.

$$R_F = R_o\,(1 + \alpha(T_2 - T_1)$$

where α = 0.00323 for copper

$$T_1 = 75°\text{ C}$$
$$T_2 = 40°\text{ C}$$
$$= \frac{0.087\Omega}{1{,}000\text{ ft}} \times [1 + (0.00323)(40°\text{ C}{-}75°\text{ C})]$$
$$= \frac{0.087\Omega}{1{,}000\text{ ft}} \times (0.88695)$$
$$\frac{0.0772\Omega}{1{,}000\text{ ft}}$$

Using starting locked rotor amps code letter F:

$$= \frac{0.077\Omega}{1{,}000\text{ ft}} \times [(2)(75\text{ ft})(421\text{ A})(0.866)]$$
$$= 4.21(4)\text{ V}$$

Using starting locked rotor amps code letter K:

$$= \frac{0.077\Omega}{1{,}000\text{ ft}} \times [(2)(75\text{ ft})(677\text{ A})(0.866)]$$
$$= 6.77(7)\text{ V}$$

Using full load amps when running:

$$= \frac{0.077\Omega}{1{,}000\text{ ft}} \times [(2)(75\text{ ft})(77\text{ A})(0.866)]$$
$$= 0.77(1)\text{ V}$$

PRACTICE EXERCISE: Provide a branch circuit for a 5 hp 230 V 3-phase motor, locked rotor code letter F and service factor 1.0 with a 50° C rise. Select the conductors, overload device, and time-delay fuse overcurrent device. Controller terminals are suitable only with copper conductors and 60° C current flow. Provide an equipment grounding conductor and use 90° C insulation on all conductors.

3.13 WELDERS

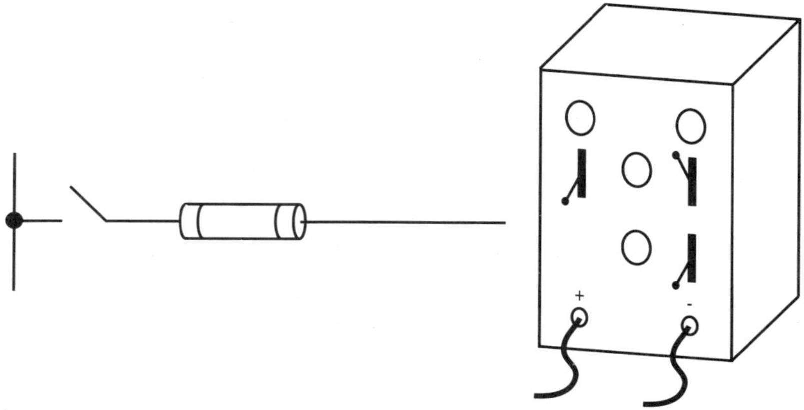

3.13.1 AC Transformer and DC Rectifier Arc Welders

EXAMPLE:

A welder has a nameplate rating of 30 amperes input and 300 amperes output with a 100% duty cycle, when supplied by 480 V three- phase using 90° C insulated conductors selected from the 60°C column.

► Conductor size: *Section 630-1(a)*
(30 A)(100%) = 30 A; use #10 AWG.

► Overcurrent protection: *Section 630-12(a)*
(30 A)(200%) = 60 A maximum fuse or circuit breaker, which is not more than 200% of the ampacity of #10 AWG.

3.13.2 Motor Generator Arc Welders

EXAMPLE:

A welder has a nameplate rating of 60% duty cycle with motor operating at 460 V and 30 amperes.

► Conductor size: *Section 630-21(a)*
(30 A)(81%) = 24.3 A; use #12 AWG.

► Overcurrent protection: *Section 630-22*
(30 A)(200%) = 60 A maximum fuse or circuit breaker, which is not more than 200% of the ampacity of #12 AWG.

3.13.3 Resistance (Spot) Welders

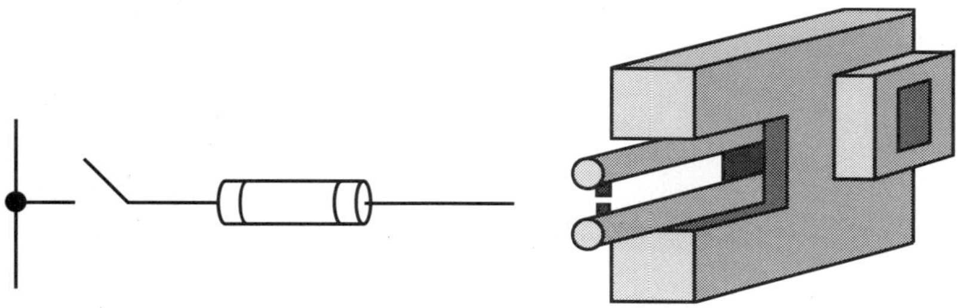

EXAMPLE:

A welder of the Article 630 Part D type has a 150 kVA transformer at 480 V single-phase, a duty cycle of 40% when manually operated and used for a specific operation.

► Conductor size: *Section 630-31*

$$\frac{(150\ \text{kVA})}{0.480\ \text{kV}} = 312.5(313)\text{A}(0.63) = 197.19(197)$$

Use 3/0 AWG.

► Transformer overcurrent protection: *Section 630-32(a)*
(313 A)(300%) = 939(900) A; use 900 A.

► Conductor overcurrent protection: *Section 630-32(b)*
(225 A)(300%) = 775(700) A maximum size fuse or circuit breaker, which is not more than 300% of the conductor ampacity but is less than the 900 A transformer protector first selected, and is the limiting factor; therefore the maximum device permitted is 700 A.

PRACTICE EXERCISE: Provide a branch circuit for a DC rectifier MIG arc welder 480 V 3-phase 30 amp rated 100% duty at 300A DC output and 87% duty at 450A DC output. It will be plug- and cord-connected for production needs.

CHAPTER 4

Calculations

4.1 DWELLING

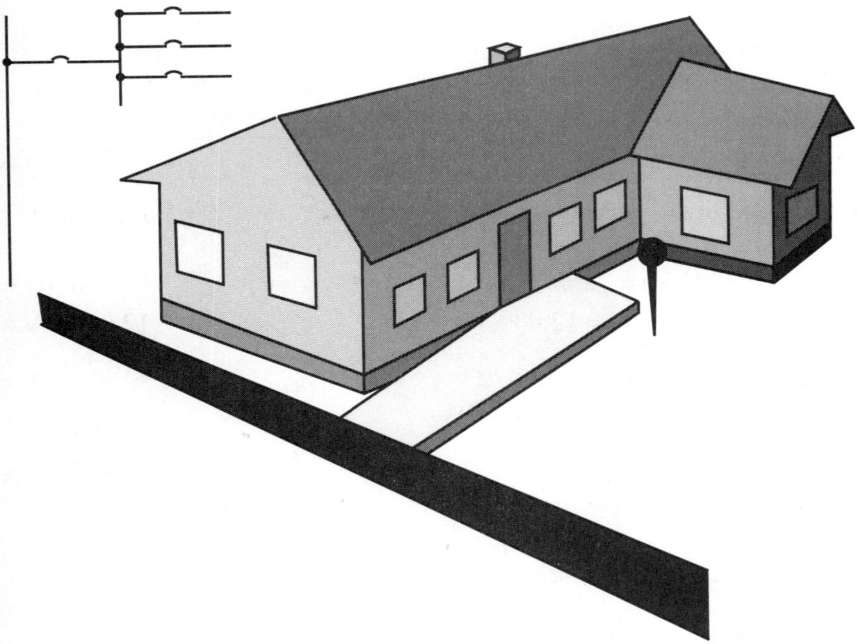

► Use *Article 220* Part B Calculation Method.

EXAMPLE:

A 1,200 ft^2 dwelling, with a 12.5 kW electric range, 12 kW individually thermostat controlled heat, 32.8 A 230 V single-phase air conditioner, 1.2 A 120 V bathroom exhaust fan, 0.96 A 120 V range exhaust fan, and a ½ hp 115 V trash compactor. There will be three circuits for the kitchen, and a laundry circuit is provided, as is a ½ hp 115 V garbage disposer.

4.1.1 Phase Conductor and Overcurrent Requirements

▶ Use *Section 220-10.*

▶ General lighting: *Section 220-3(b)*

$$(1{,}200 \text{ ft}^2)\,\frac{(3 \text{ VA})}{\text{ft}^2} = 3{,}600 \text{ VA}$$

▶ Small appliance and laundry circuits: *Section 220-16(a)* and *(b)*

Three circuits @ 1,500 VA	= 4,500 VA	
Laundry circuit	= 1,500 VA	
Subtotal	9,600 VA	

▶ Demand factors: *Section 220-11*

Less 3,000 kW @100%	−3,000 VA = 3,000.0 VA	
Less remainder @ 35%	6,600 VA = 2,310.0 VA	
Subtotal		5,310.0 VA

▶ Air conditioner vs. electric heat: *Sections 220-14, 220-15*, Ex. 2, and *Section 220-21*

(32.8 A)(240 V) = 7,544 VA	=	0.0 VA
(12,000 W)(100%) = 12,000 VA	=	12,000.0 VA

▶ Electric range: *Section 220-19*

12.5 kW − 12.0 kW = 0.5(1) kW

$$8 \text{ kW} + (8 \text{ kW})\,\frac{(5\%)}{\text{kW}}\,(1 \text{ kW}) \quad = \quad 8{,}400.0 \text{ VA}$$

▶ Electric dryer: *Section 220-18*

4.5 kW; use	=	5,000.0 VA

▶ Miscellaneous appliance loads: *Section 220-17*

Fan @ (1.2 A)(120 V)	= 144.0 VA	
Fan @ (0.96 A)(120 V)	= 115.2 VA	
Disposer @ (9.8 A)(120 V)	= 1,176.0 VA	
Trash compactor @ (16 A)(120 V)	= 1,920.0 VA	
Water heater	= 5,000.0 VA	
Demand factor	(75%)(8,355.2 VA) =	6,266.4 VA
Subtotal		36,996.4 VA

► *Section 220-14.*

Plus 25% largest motor	=	480.0 VA
Total		37,456.4 VA

$$\frac{(37{,}456.4\ \text{VA})}{240\ \text{V}} = 156.06(156)\ \text{A}$$

► Use #1/0 AWG with 175 A service rating (*Section 240-3* and Ex. 4).

4.1.2 Neutral Conductor Requirements

Use *Section 220-22.*

Lighting, small appliance and laundry circuits		=	5,310.0 VA
Air conditioner		=	0.0 VA
Electric heat		=	0.0 VA
Electric range @ (8,400 VA)(70%)			
Electric dryer @ (5,000 VA)(70%)		=	3,500.0 VA
Appliance loads: *Section 220-17*			
Fan	= 144.0 VA		
Fan	= 115.2 VA		
Disposer	= 1,176.0 VA		
Water heater	= 0.0 VA		
Demand factor	(75%)(3,355.2 VA)	=	2,516.4 VA
Subtotal			17,172.4 VA
Plus 25% largest motor			294.0 VA
Total			17,466.4 VA

$$\frac{(17{,}466.4\ \text{VA})}{(2)(120\ \text{V})} = 72.7(73)\ \text{VA}$$

Although *Tables* 310-16 through 310-19 permit a minimum of #4 AWG, which also satisfies *Section 250-23(b)*, Note 3 to those tables permits no less than #2 AWG.

4.1.3 Grounding Electrode Conductor

► Use *Section* and *Table 250-94*, select minimum #6 AWG.

4.1.4 Raceway Size

► Use *Section 300-17.*

XHHW insulation: (2)(0.1893) = 0.3786
(1)(0.1182) = 0.1182
0.4968

Use 1-½ inch.

PRACTICE EXERCISE: Determine the minimum service entrance conductor sizes and service entrance disconnect rating for a single family dwelling with 1,400 ft^2, 12 kW electric range, provide for a 4 kW electric dryer, a heat pump with a nameplate requiring 32.4 A minimum circuit capacity and 40 A maximum overcurrent protection and 15 kW central electric furnace, a ¾ hp garbage disposer, and a 240 V water heater with 2 heating elements at 5 kW each with a double-pole double-throw thermostat. Use *Article 220* Part B method for the calculations.

4.2 MORE DWELLING

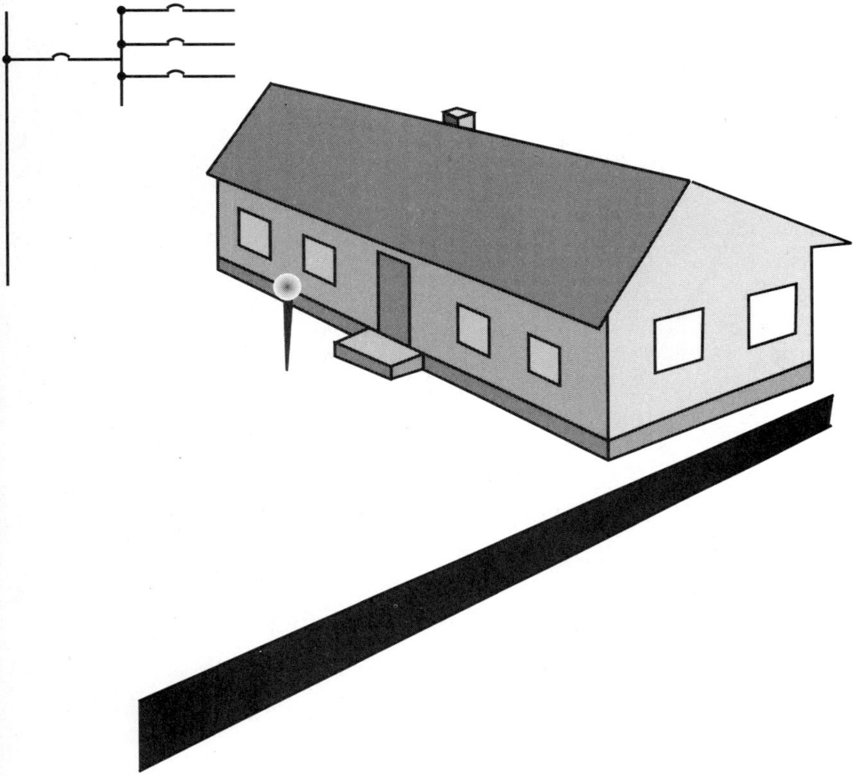

EXAMPLE:
A 1,200 ft^2 dwelling, with a 12.5 kW electric range, 12 kW individually thermostat controlled heat, 32.8 A 230 V single-phase air conditioner, 1.2 A 120 V bathroom exhaust fan, 0.96 A 120 V range exhaust fan, and a ½ hp 115 V trash compactor. There will be three circuits for the kitchen, and a laundry circuit is provided.

► Use *Article 220* Part C Method and *Table 220-3(a)*.

4.2.1 Phase Conductor and Overcurrent Requirements

Air conditioning vs. electric heat:
(32.8 A)(240 V) = 7,872 VA = 7,872.0 VA
(12,000 kW)(40%) = 4,800 VA = 0.0 VA

► *Section 220-3(b)*

(1,500 VA)(3 sm appl ckts) = 4,500.0 VA

(1,500 VA)(1 laundry ckt) = 1,500.0 VA

$(1{,}200\ \text{ft}^2)\dfrac{(3\ \text{VA})}{\text{ft}^2}$ = 3,600.0 VA

Fan @ (1.2 A)(120 V) = 144.0 VA

Fan @ (0.96 A)(120 V) = 115.2 VA

Dryer @ 4.5 kW = 4,500.0 VA

Water heater @ 5 kW = 5,000.0 VA

Trash compactor
@ (16 A)(120 V) = 1,920.0 VA

Elec range @ 12.5 kW = 12,500.0 VA

Subtotal 33,779.2 VA

Less first 10,000 VA − 10,000.0 VA =10,000.0 VA

Remainder @ (40%)(23,779.2 VA) = 9,511.6 VA

Total 27,383.6 VA

$\dfrac{(27{,}383.6\ \text{VA})}{240\ \text{V}} = 114.1(114)\ \text{A}$

► Phase conductor: (*Table 310-16* Note 3).
Use minimum #2 AWG and 125 A service rating.

4.2.2 Neutral Conductor

► Use *Section 220-22*:

Lighting, small appliance
and laundry circuits = 5,310.0 VA

Air conditioner = 0.0 VA

Electric heat = 0.0 VA

Electric range @ (8,400 VA)(70%) = 5,858.0 VA

Electric dryer @ (5,000 VA)(70%) = 3,500.0 VA

► Appliance loads: *Section 220 – 117*

Fan = 1 44.0 VA

Fan = 115.2 VA

Disposer = 1,176.0 VA

Water heater 0.0 VA

Demand factor (75%)(3,355.2 VA) = 2,516.4 VA

Subtotal 17,172.4 VA

Plus 25% largest motor	294.0 VA
Total	17,466.4 VA

$$\frac{(17{,}466.4\ \text{VA})}{(2)(120\ \text{V})} = 72.7(73)\ \text{VA}$$

► *Tables 310-16* through *19* Note 3 require no less than #4 AWG, and also satisfy *Section 250-23(b)*. However, unless the service entrance panelboard and all installed circuit breakers have a 75°C rating, it would have to be increased to #3 AWG, per *Section 110-14(c)*.

4.2.3 Grounding Electrode Conductor

► Use *Section* and *Table 250-94*, select minimum #8 AWG.

4.2.4 Raceway Size

Use *Section 300-17*:

XHHW insulation: (2)(0.1182) = 0.2364
(1)(0.0845) = 0.0845
0.3209

Minimum 1 inch.

PRACTICE EXERCISE: Determine the minimum service entrance conductor sizes and service entrance disconnect rating for a single family dwelling with 1,400 ft^2, 12 kW electric range, provide for a 4 kW electric dryer, a heat pump with a nameplate requiring 32.4 A minimum circuit capacity and 40 A maximum overcurrent protection and 15 kW central electric furnace, a $\frac{3}{4}$ hp garbage disposer, and a 240 V water heater with 2 heating elements at 5 kW each with a double-pole double-throw thermostat. Use *Article 220* Part C method for the calculations.

4.3 OFFICE BUILDING

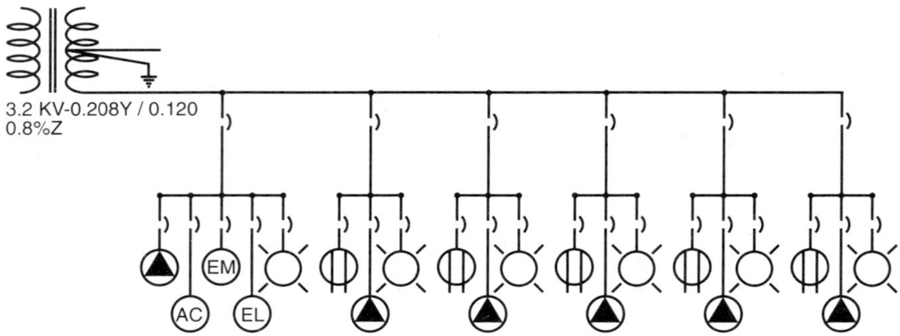

EXAMPLE:

Consider a five-story 40 ft by 60 ft structure having a 10 ft by 10 ft stairwell area and 10 ft by 10 ft elevator area provided for each floor. Assume a continuous load value for lighting, 75 receptacle outlets per office floor, with both subject to harmonic conditions. 15 kW of infinitely controlled electric heat is also provided for each floor with 7.5 kW for the basement storage area.

The passenger elevator car requires a 20-ampere branch circuit and a 5 hp 3-phase hydraulic motor.

The maximum load desired for the receptacle and lighting circuit is 50% of their circuit rating, with all terminals rated for 75° C ampacity except for the motor circuits, which are limited to 60° C. The utility will provide a 150 kVA 208Y/120 V 3-phase source having 0.8% impedance, and the cut length of the service entrance conductors will be 30 feet. All conductors will have 90° C insulation.

▶ Use *Section 110-14.*

4.3.1 Dimensions

Total square foot area of each floor is:
$(40 \text{ ft} \times 60 \text{ ft}) = 2{,}400 \text{ ft}^2$

Total square foot area of the stairway and elevator at each floor is:
$(10\text{ ft} \times 10\text{ ft}) = 100\text{ ft}^2$

Net office area per floor is:
$(2{,}400\text{ ft} - 100\text{ ft} - 100\text{ ft}) = 2{,}200\text{ ft}^2$

Net storage area in the basement is also $2{,}200\text{ ft}^2$.

4.3.2 General Lighting Load per Floor

► Use *Section* and *Table 220-3(b)*:

$$(2{,}200\text{ ft}^2)\ \frac{(3.5\text{ VA})}{\text{ft}^2}\ (1.25) = 9{,}625.0\text{ VA; use \#12 AWG.}$$

General lighting load per pole:

$$\frac{(2{,}200\text{ ft}^2)(3.5\ \text{VA})(125\%)}{(120\text{ V})(\text{ft}^2)(10\ \text{A/ pole})} = 8.02(9)\text{ poles}$$

$$\frac{9{,}625.0\text{ VA}}{9\text{ poles}} = \frac{1{,}069.4\text{ VA}}{\frac{(120\text{ V})}{\text{pole}}\ (\text{pole})} = 8.9(9)\text{ A}$$

4.3.3 Receptacle Load per Office Floor

$$(75\text{ receptacles})\ \frac{(180\text{ VA})}{\text{receptacles}} = 13{,}500\text{ VA; use \#12 AWG}$$

Receptacles per pole:

$$\frac{(13{,}500\text{ VA})}{(3\text{-phases})\ \frac{(120\text{ v})}{\text{phase}}\ (50\%)(20\text{ A})} = 3.75(4)$$

4.3.4 Elevator Passenger Car Branch Circuit Load

► Use *Sections 620-13(a)* and *620-61(b)(1)*:
(20 A)(80%) = 16.0 A; use #12 AWG and 1,920 W elevator motor 5 hp @ 208 V.
(16.7 A)(140%) = 23.38(23) A; use #12 AWG.
(16.7 A)(175%) = 29.2(29) A; use 25 A TD fuse, or
(16.7 A)(250%) = 41.7(42) A; use 40 A ckt breaker

4.3.5 Air Conditioning Motor

Hermetic motor 25 hp @ 208 V. (*Sections 440-32 and 440-22(a)*):
(74.8 A)(125%) = 93.5(94) A; use #2 AWG.
(74.8 A)(175%) = 130.9(131) A; use 125 A fuse or circuit breaker.

4.3.6 Electric Heat per Office Floor

► Use *Section 424-3*:

$$15 \text{ kW} = \frac{(15{,}000 \text{ VA})(1.25)}{\frac{(120 \text{ V})}{\text{phase}}(3\text{-phase})} = 52.0 \text{ A};$$

Use #6 AWG and 60 A fuse or circuit breaker.

4.3.7 Office Feeder Phase Conductor Size

► Use *Section 220-10(b)*:

General lighting: $(2{,}220 \text{ ft}^2)\frac{(3.5 \text{ VA})}{\text{ft}^2} =$ 7,700 VA

Receptacles:

► Use *Section 220-13*:

	13,500 VA	
Less	(− 10,000 VA)	= 10,000 VA
Net	(50%)(3,500 VA)	= 1,750 VA
Electric heat		= 15,000 VA
Subtotal		= 34,450 VA
Plus 25% harmonic allowance		= 8,612 VA
Total		43,062 VA

$$\frac{(43{,}062 \text{ VA})}{(3)(120 \text{ V})} = 119.6(120) \text{ VA; use minimum \#1 AWG.}$$

4.3.8 Neutral Conductor Size

► Use *Section 220-22*:

General lighting	= 7,700 VA
Receptacles	= 11,750 VA
Electric heat	= 0 VA
Subtotal	= 19,450 VA

Plus 0.73 harmonic adder	= 14,199 VA
Total	= 33,649 VA

$$\frac{(33{,}649\text{ VA})}{(3)(120\text{ V})} = 93.4(94)\text{ A; use \#3 AWG.}$$

► Minimum feeder overcurrent protection: *Section 220-10(a)*

General lighting	= 7,700 VA
Receptacles	= 11,750 VA
Subtotal	= 19,450 VA
Plus 25% continuous load allowance	= 4,863 VA
Subtotal	= 24,313 VA
Electric heat	= 15,000 VA
Total	= 39,313 VA

$$\frac{39{,}313\text{ VA}}{(3)(120\text{ V})} = 109.2(109)\text{ A; use 110 A fuse or circuit breaker.}$$

► Minimum feeder equipment grounding conductor size:
Use *Section* and *Table 250-95*, select minimum #6 AWG.
#6 AWG.

► Minimum raceway size: *Section 300-17*

THHN insulation: (3)(0.1590) = 0.6360
(1)(0.0995) = 0.0995
(1)(0.0519) = 0.0519
0.7874

Use 1-½ inch.

4.3.9 Basement Calculations

► Basement phase conductors: *Section 220-10(b)*
General lighting load

$$(2{,}200\text{ ft}^2)\ \frac{(\frac{1}{4}\text{ VA})}{\text{ft}^2} = 550\text{ VA}$$

Stairwell lighting

$$\frac{(100\ \text{ft}^2)(\tfrac{1}{5}\ \text{VA})}{(\text{floors})(\text{ft}^2\ )}(6\ \text{floors}) = 120\ \text{VA}$$

Subtotal	=	670 VA
Plus 25% harmonic allowance	=	168 VA
Elevator passenger car	=	1,920 VA
Elevator motor (16.7 A)(360 V)	=	6,012 VA

► Hermetic motor: *Sections 220-21* and *430-25*

(74.8 A)(360 V)	=	26,928 VA
Plus 25% largest motor	=	6,732 VA
Electric heat @ 7.5 kW	=	0 VA
Total	=	42,430 VA

$$\frac{(42{,}430\ \text{VA})}{(3)(120\ \text{V})} = 117.9(118)\ \text{A; use \#1 AWG.}$$

► Basement neutral conductor: *Section 220-22*

General lighting (670 VA)(1.73)	=	1,159 VA
Elevator passenger car	=	1,920 VA
Elevator motor	=	0 VA
Hermetic motor	=	0 VA
Total	=	3,079 VA

$$\frac{(3{,}079)}{((3)(120\ \text{V})} = 8.6(9)\ \text{A}$$

► Use *Section 215-2(a)* and select #10 AWG.

► Basement non-motor load overcurrent protection: *Section 430-63*

General lighting	=	670 VA
Stairwell lighting	=	300 VA
Elevator passenger car	=	1,920 VA
Electric heat	=	0 VA
Subtotal	=	2,890 VA
Plus 25% continuous duty	=	723 VA
Subtotal	=	3,613 VA

$$\frac{(3{,}613\ \text{VA})}{(3)(120\ \text{V})} = 10.0\ \text{A}$$

► Use *Section 215-2(a)* and select 30 A and #10 AWG.

► Basement motor feeder protection: *Section 240-3(f)*

Hermetic motor	=	125.0 A
Elevator motor	=	16.7 A
Total	=	141.7 A; use 125 A

Combined basement protection:
125 A + 10 A = 135 A; use 150 A.

► Minimum equipment grounding conductor: *Section 250-95*, #6 AWG

► Minimum basement feeder raceway: *Section 300-17*

THHN insulation.	(3)(0.1590) =	0.4770
	(1)(0.0216) =	0.0216
	(1)(0.0625) =	0.0625
		0.5611

Use 1-¼ inch.

4.3.10 Total Building Calculations

► Building service phase conductors: *Section 220-10(a)*

General lighting (5)(7,700 VA)		= 38,500 VA		
Stairwell lighting		= 120 VA		
Receptacle load:				
(5)(13,500 VA)	= 67,500 VA			
Less	−10,000 VA	= 10,000 VA		
Subtotal (50%)	(57,500 VA)	= 28,750 VA		
Subtotal			=	77,370 VA
Plus 25% harmonic allowance			=	19,342 VA
Elevator car			=	1,920 VA
Hermetic motor		= 0 VA		
Elevator motor		= 6,012 VA		
Plus 25%		= 1,503 VA		
Subtotal			=	7,515 VA

Electric heat:		
Office areas (5)(15,000 kW)	= 75,000	
Basement area	= 7,500	
Subtotal		= 82,500 VA
Total		= 188,647 VA

$\frac{(188{,}647\ \text{VA})}{(3)(120\ \text{V})} = 524$ A; use minimum 2-300 kcmil per phase in separate raceways.

► Building service neutral conductor: *Section 220-22*

General lighting and receptacles	= 77,370 VA
Plus 73% for harmonic allowance	= 56,480 VA
Elevator car	= 1,920 VA
Hermetic motor	= 0 VA
Elevator motor	= 0 VA
Electric heat	= 0 VA
Total	= 135,770 VA

$\frac{(135{,}770\ \text{VA})}{(3)(120\ \text{V})} = 377$ A; use 2 3/0 per raceway.

► Minimum building overload protection: *Section 230-90*

Total non-motor load:

$\frac{[(79{,}290\ \text{VA})(125\%) + 82{,}500\ \text{VA}]}{(3)(120\ \text{V})}$	= 504 A
Elevator time delay fuse	= 15 A
Total	= 519 A

► Use 600 A fuse: *Section 240-3* and Ex. 4.

4.3.11 Above-Ground Service Entrance Conductors

Use minimum conductor ampacity with the major portion of the neutral load subject to harmonic currents. Consider two parallelled 300 kcmil conductors per phase with XHHW insulation in two raceways for a total of 4 conductors in each.

► Use *Section 310-15(a)*:

(2)(285 A) = 570 A(80%) = 456 A, which is less than the 524 A calculated load.

Increase size to two parallelled 400 kcmil conductors with XHHW insulation.
(2)(335 A) = 670 A(80%) = 536 A, which is greater than the 524 A required and is acceptable.
For the neutral, consider the two parallelled 3/0 AWG XHHW insulation.
(2)(200 A) = 400 A(80%) = 320 A, which is less than the 377 A calculated.
Increase size to two parallelled 250 kcmil conductors with XHHW insulation.
(2)(255 A) = 510 A(80%) = 408 A which is greater than the 377 A required and is acceptable.

► (This will also satisfy *Section 250-23(b)*.)

► Minimum raceway: *Section 300-17*

$$\begin{aligned}(3)(0.5931) &= 1.7793\\ (1)(0.4026) &= \underline{0.4026}\\ &\ 2.1819\end{aligned}$$

Use 3 inch.

► Grounding electrode conductor: Use *Section* and *Table 250-94* and select minimum #2/0 AWG.

4.3.12 Underground Service Entrance Conductors

Double-check service run underground from the pad-mounted transformer using information in Appendix B. Burial depth is 6 ft, backfill material is sandy, and load factor is determined by comparing the transformer full load current rating and the calculated service overload protection.

► Use *Section 310-15(b)*; also use *Table B-310-7* and Figure B-310-1:

$$\frac{I_2}{I_1} \text{ ratio is } \frac{\text{R(ho) 120 LF 100}}{\text{R(ho) 60 LF 50}} = \frac{361\text{ A}}{442\text{ A}} = .81\text{; use }.80$$

Sandy soil, assume R = 120

$$\text{LF} = \frac{416.7\text{ A}}{600\text{ A}} = .69\text{; use }70$$

► Use Figure B-310-1.
Read across the bottom to R(ho) = 120, then upward to intersect LF = 70 line from left-hand side. Follow, parallelling the curved line up-

ward and to the left slightly to line showing where LF = 100, and the resultant indicates 58 (use 60). Read vertically upward to where it intersects the $\frac{I_2}{I_1}$ ratio = .80 curved line from the right-hand side.

Reading then straight across to the left the resultant F = .87 for a single duct (raceway) and *Section B-310-15(b)(5)(1)* further reduces it by .88, or (.87)(.88) = (.7656)(442)(2) = 676.8(677) A. Note 8(a) reduces it further to (80%)(677 A) = 541.6(542 A).

With increased burial depth and using *Section B-310-15(b)(3)(2)* criteria modifications,

$$\left(100\% - (6\text{ft} - 2\tfrac{1}{2}\text{ ft})\frac{(6\%)}{\text{ft}}\right) = 79\%$$

(79%)(542 A) = 428.2(428) A, which is less than the 524 A calculated and indicates overheating may occur. Increasing the size to 500 kcmil results in an $\frac{I_2}{I_1}$ ratio of .81 (use .80) with the F remaining at .87, and the resultant (503 A)(2)(.7656)(80%) = 616.2 A is greater than the calculated 524 A. The resultant underground neutral ampacity is 409.1(409) A, with no increase necessary for it, the grounding electrode conductor or raceway size.

▶ Calculated fault currents: *Sections 110-9, 110-10,* and *230-65*

$$\frac{150\text{ kVA}}{(3)(.012\text{kV})(0.0008)} = 52{,}083(52{,}000)\text{ A}$$

Impedance of the underground parallelled phase conductors at 20° C and 30 ft long:

$$\begin{aligned} Z &= (0.050\Omega)[1 + 0.00323(20-75)] \\ &= (0.050\Omega)[1 + (-0.17765)] \\ &= (0.050\Omega)(0.82235) \\ &= \frac{(0.041\Omega)(30\text{ ft})}{(1{,}000\text{ ft})(2)} \\ &= 0.000615 \end{aligned}$$

Impedance of the underground parallelled neutral conductors at 20° C and 30 ft:

$$
\begin{aligned}
Z &= (0.073\Omega)[1 + 0.00323(20 - 75)] \\
&= (0.073\Omega)(1 - 0.1615) \\
&= (0.073\Omega)(0.8385) \\
&= \frac{(0.061\Omega)(30 \text{ ft})}{(1{,}000 \text{ ft})(2)} \\
&= 0.000915
\end{aligned}
$$

Impedance of the source:

$$
\begin{aligned}
Z &= \frac{120 \text{ V}}{52{,}000 \text{ A}} \\
&= 0.0023\Omega
\end{aligned}
$$

$$
\begin{aligned}
I &= \frac{120 \text{ V}}{(0.000615\Omega) + (0.000915\Omega) + (0.0023\Omega)} \\
&= \frac{120 \text{ V}}{.003830\Omega} \\
&= 31{,}331 \text{ A}
\end{aligned}
$$

6 cycle withstand rating for 1,000,000 cmil exceeds 100,000 A.
6 cycle withstand rating for 500,000 cmil exceeds 80,000 A.
6 cycle withstand rating for 250,000 cmil exceeds 40,000 A.

If the fault persists for a longer period, cable damage may occur; steps should be taken to minimize this. Many other methods exist for calculating fault currents, one of which is shown in Chapter 1. They vary in degree of difficulty. It is recommended the student review their concepts.

PRACTICE EXERCISE: Determine the service requirements for a single-story 60 ft by 60 ft office building with non-linear lighting, 60 duplex receptacles supplying computers and similar loads, a 25 hp blower motor for a gas furnance, a 30 hp standard motor for the building air conditioning system, and a 5 kW hot water heater. The power supply is from a pad-mounted transformer 75 kVA 208Y/120 V, 1.67% impedance, with the lateral conductors in metallic raceway 48 inches burial depth, with a cut length of 25 feet. Conductor insulation is 90° C dry, 75° C damp or wet rated. Determine conductor size, service overload protection, raceway size, grounded conductor, and grounding electrode conductor.

NOTES

NOTES

NOTES

NOTES

NOTES